D1264573

Guide to Electric Power Generation

3rd Edition

Guide to Electric Power Generation

3rd Edition

A.J. Pansini
K.D. Smalling

THE FAIRMONT PRESS, INC.

CRC Press
Taylor & Francis Group

Library of Congress Cataloging-in-Publication Data

Pansini, Anthony J.
     Guide to electric power generation / A.J. Pansini, K.D. Smalling.--3rd ed.
     p. cm.
   Includes index.
   ISBN 0-88173-524-8 (print) -- ISBN 0-88173-525-6 (electronic)
     1. Electric power production. 2. Electric power plants. I. Smalling,
Kenneth D. 1927-
   II. Title.

   TK1001 .P35 2005
   621.31--dc22

                            2005049470

Published by The Fairmont Press, Inc.
700 Indian Trail
Lilburn, GA 30047
tel: 770-925-9388; fax: 770-381-9865
http://www.fairmontpress.com

Distributed by Taylor & Francis Ltd.
6000 Broken Sound Parkway NW, Suite 300
Boca Raton, FL 33487, USA
E-mail: orders@crcpress.com

Distributed by Taylor & Francis Ltd.
23-25 Blades Court
Deodar Road
London SW15 2NU, UK
E-mail: uk.tandf@thomsonpublishingservices.co.uk

Printed in the United States of America
10 9 8 7 6 5 4 3 2 1

0-88173-524-8 (The Fairmont Press, Inc.)
0-8493-9511-9 (Taylor & Francis Ltd.)

Contents

Preface

Like water, food, and air, electrical energy has become an integral part of daily personal and business lives. People have become so accustomed to flicking a switch and having instant light, action, or communication that little thought is given to the process that produces this electrical energy or how it gets to where it is used. It is unique in that practically all that is produced is not stored but used instantly in the quantities that are needed. For alternatives to electrical energy, one must go back to the days of gas lamps, oil lamps, candles, and steam- or water-powered mechanical devices—and work days or leisure time that was limited to daylight hours for the most part.

Where does this vital electrical energy come from and how does it get to its users? This book covers only the how, when and where electrical energy is produced. Other texts cover how it is delivered to the consumer. The operations of an electric system, like other enterprises may be divided into three areas:

Electric Generation (Manufacturing)
Electric Transmission (Wholesale Delivery)
Electric Distribution (Retailing)

The electric utility is the basic supplier of electrical energy and is perhaps unique in that almost everyone does business with it and is universally dependent on its product. Many people are unaware that a utility is a business enterprise and must meet costs or exceed them to survive. Unlike other enterprises producing commodities or services, it is obligated to have electrical energy available to meet all the customer demands when they are needed, and its prices are not entirely under its control.

The regulation of utilities by government agencies leads to the perception that utilities are in fact monopolies. People have alternatives in almost every other product they use such as choosing various modes of travel—auto, train or plane. People can use gas, oil or coal directly for their own energy needs or use them to generate their own electrical energy. Indeed some people today use sunlight or windpower to supplement their electrical energy needs. The point is that electrical energy supply from an electric system is usually much more convenient and economical than producing it individually. Some larger manufacturing firms find it feasible to provide their own electrical energy by using their waste energy (cogeneration) or having their own individual power plants. In some cases legislation makes it mandatory to purchase the excess energy from these sources at rates generally higher than what the utility can produce it for.

The fact remains that utilities must pay for the materials, labor and capital they require and pay taxes just like other businesses. In obtaining these commodities necessary to every business, utilities must compete for them at prices generally dictated by the market place, while the prices charged for the product produced—electrical energy—are limited by government agencies.

Since our first edition, electric systems have been moving toward deregulation in which both consumer and supplier will be doing business in a free market—which has no direct effect on the material contained in the accompanying text.

The problems faced with producing electrical energy under these conditions are described in this text in terms which general management and non-utility persons can understand. Semi-technical description in some detail is also included for those wishing to delve more deeply into the subject.

None of the presentations is intended as an engineering treatise, but they are designed to be informative, educational, and adequately illustrated. The text is designed as an educational and training resource for people in all walks of life who may be less acquainted with the subject.

Any errors, accidental or otherwise, are attributed only to us.

Acknowledgment is made of the important contributions by Messrs. H.M. Jalonack, A.C. Seale, the staff of Fairmont Press and many others to all of whom we give our deep appreciation and gratitude.

Also, and not the least, we are grateful for the encouragement and patience extended to us by our families.

Waco, Texas   *Anthony J. Pansini*
Northport, N.Y.   *Kenneth D. Smalling*
1993/2001

Preface
To the Third Edition

The twentieth Century ended with more of the demand for electricity being met by small units known as Distributed Generation and by cogeneration rather than by the installation of large centrally located generating plants. Although this may appear to be a throwback to earlier times when enterprises used windmills and small hydro plants for their power requirements, and a bit later with these converted to electric operation, then making such "left over" power available to the surrounding communities, the return to local and individual supply (cogeneration) may actually be pointing in the direction of future methods of supply. Will the end of the Twenty First Century see individual generation directly from a small unit, perhaps the rays from a few grains of radioactive or other material impinging on voltaic sensitive materials, all safely controlled ensconced in a unit that takes the place of the electric meter?

There are many advantages to this mode of supply. Reliability may approach 100 percent. When operated in conjunction with Green Power systems, supplying one consumer tends to make security problems disappear and improvements in efficiency and economy may be expected. Transmission and distribution systems, as we know them, may become obsolete

The output of such systems will probably be direct current, now showing signs of greater consideration associated with Green Power units: fuel cells, solar power and others, all provide direct current. Insulation requirements are lower, synchronizing problems disappear, and practical storage of power is enhanced—all pointing to the greater employment of direct current utilization.

The current trend toward Distributed Generation, employing primary voltages and dependence on maintenance standards being followed by "lay" personnel, pose safety threats that do not occur with the systems envisioned above. With education beginning in the lower grades about the greater ownership of such facilities by the general public, all tend to safer and foolproof service.

The Twenty First Century should prove exciting!

Introduction

The last two decades of the Twentieth Century saw a distinct de-cline in the installation of new generation capacity for electric power in the United States. With fewer units being built while older plants were being retired (some actually demolished), the margin of availability compared to the ever increasing demand for electricity indicated the approach of a shortage with all of its associated problems. (This became reality for consumers in California who experienced rolling blackouts and markedly high energy costs.) Perhaps spurred by the deregulation initiated for regulated investor utilities, an effort to reverse this trend began at the end of the century to restore the vital position of power generation in the new millennium, described in Chapter One. The appar-ent decline in constructing new generation may be explained by several factors:

- The decline of nuclear generation in the U.S. because of adverse public opinion, and soaring costs caused by the increasing com-plexity of requirements promulgated by federal agencies.

- The introduction of stringent rules for emissions by the Clean Air Act and other local regulations.

- The reluctance of regulated utilities to risk capital expenditures in the face of deregulation and divestiture of generation assets, as well as uncertainty of final costs from changing government regulations.

- The endeavors to meet electric demands through load manage-ment, conservation, cogeneration (refer to Figure I-1), distributed generation, and green power (fuel cells, wind solar, micro turbine, etc.) (refer to Figure I-2).

The choice of fuels for new plants presents problems. Fossil fuels still predominate but are more than ever affected by environmental and political considerations. Natural gas, the preferred "clean fuel", is in short supply while new explorations and drilling are subject to many non-technical restrictions. The same comment applies to oil, although the supply, while more ample, is controlled by foreign interests that fix prices. The abundance of coal, while fostering stable prices, can no longer be burned in its natural state but must be first processed for cleaner burning. While many other nations (e.g. France) obtain much of their electrical energy requirements from nuclear power, the United States that pioneered this type plant sells abroad but does not compete in this country.

It is also evident that new transmission lines to bring new sources of energy to load centers will be required (but are not presently being built). Such lines now become the weak link in the chain of deregulated supply. Notoriously, such lines are built in out-of-the-way places for environmental and economic reasons and are subject to the vagaries of nature and man (including vandals and saboteurs). Who builds them, owns and operates them, is a critical problem that must be solved in the immediate future.

Implementation of other power sources such as fuel cells, solar panels, wind generators, etc., needs development—from expensive experimental to large-scale economically reliable application. Hydropower may find greater application, but its constancy, like wind and sun, is subject to nature's whims.

The new millennium, with the changing methods of electric supply brought about by deregulation, may see some alleviation in the problems associated with generating plants. It will also see new challenges that, of a certainty, will be met by the proven ingenuity and industry of our innovators and engineers, the caretakers of our technology.

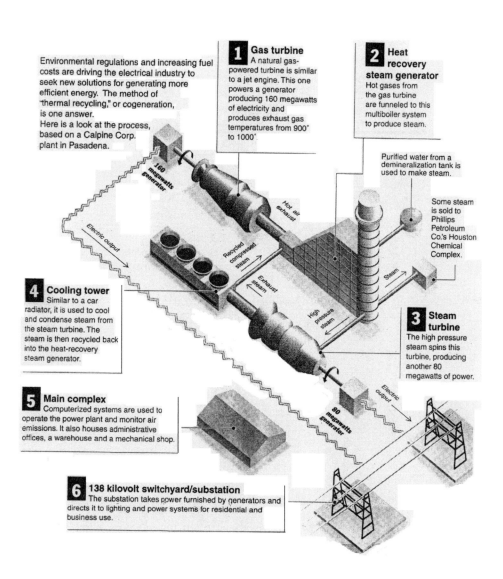

Environmental regulations and increasing fuel costs are driving the electrical industry to seek new solutions for generating more efficient energy. The method of "thermal recycling," or cogeneration, is one answer.
Here is a look at the process, based on a Calpine Corp. plant in Pasadena.

1 Gas turbine
A natural gas-powered turbine is similar to a jet engine. This one powers a generator producing 160 megawatts of electricity and produces exhaust gas temperatures from 900° to 1000°

2 Heat recovery steam generator
Hot gases from the gas turbine are funneled to this multiboiler system to produce steam.

Purified water from a demineralization tank is used to make steam.

Some steam is sold to Phillips Petroleum Co.'s Houston Chemical Complex.

4 Cooling tower
Similar to a car radiator, it is used to cool and condense steam from the steam turbine. The steam is then recycled back into the heat-recovery steam generator.

3 Steam turbine
The high pressure steam spins this turbine, producing another 80 megawatts of power.

5 Main complex
Computerized systems are used to operate the power plant and monitor air emissions. It also houses administrative offices, a warehouse and a mechanical shop.

6 138 kilovolt switchyard/substation
The substation takes power furnished by generators and directs it to lighting and power systems for residential and business use.

Figure I-1a. Cogeneration System Overview
(Courtesy Austin American Statesman)

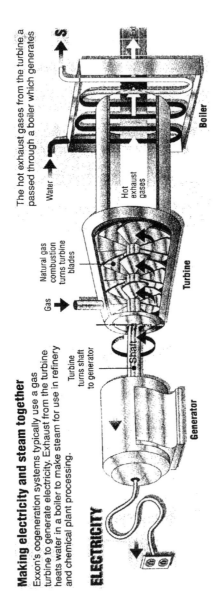

Making electricity and steam together

Exxon's cogeneration systems typically use a gas turbine to generate electricity. Exhaust from the turbine heats water in a boiler to make steam for use in refinery and chemical plant processing.

The hot exhaust gases from the turbine a passed through a boiler which generates

ELECTRICITY

Turbine turns shaft to generator

Generator

Shaft

Gas

Natural gas combustion turns turbine blades

Turbine

Hot exhaust gases

Water

Boiler

Figure I-1b. Cogeneration System with Gas Turbine (*Courtesy Exxon Corp.*)

(Top) Figure I-2a. Roof-top photovoltaic panels will play a key role in on-site power generation. The natatorium of the Georgia Institute of Technology in Atlanta uses 32,750 square feet of solar panels. (Photo: Solar Design Associates)

(Left) Figure I-2b. Wind, an important renewable source of power, may be combined in a hybrid system with a diesel backup.

(Courtesy Pure Power, Supplement to Consulting Engineering)

Figure I-2c. Microturbines can be run on any fuel, but natural gas is the fuel of choice. (Photo, Capstone Turbine) (*Courtesy Pure Power, Supplement to Consulting Engineering*)

Chapter 1

Planning and Development Of Electric Power Stations

HISTORICAL DEVELOPMENT

With the dawn of a new era in which the electric incandescent light replaced oil lamps and candles, sources of electrical energy had to be found and developed. Gas light companies were giving way to geographically small electric companies. For instance on Long Island, New York, a company called "Babylon Electric Light Company" was formed in 1886. It would surprise many LI residents today that the low level waterfall on Sumpwam's Creek in Babylon was used to light up eight stores and three street lights and that the dam still exists. Similar examples can be cited for other communities throughout the country. Most small electric companies started out using hydropower or steam engines to generate their electrical energy.

As the innovation caught on and the electrical energy requirements grew from the use of lights and electrically driven equipment, so did the growth of electric power generators. The size of generators grew from a few hundred watts to thousands of kilowatts. New sources of fuel needed to power generators led to coal, oil and gas fired boilers. New ways of transmitting electric energy for some distance was found and led to larger central stations instead of the small local area stations. As AC (alternating current) transmission developed to permit sending power over longer distances, the early small electric companies consolidated their territories and started to interconnect their systems. Planning and development of these early generating stations were not hindered by environmental restrictions or government regulations. Their main concern was raising of enough capital to build the stations and selecting the best site for the fuel to be used and the load to be served. The 1990's introduced deregulation, one result of which was the divesting of some utility generation to non-utility generation or energy companies. New generation added was mostly in the form of combustion turbine units previously used by utilities for peak loads, and not lower energy cost units such as steam turbines or hydro generators.

GROWTH OF ELECTRIC USAGE

While the growth of electric usage proceeded at a fairly steady pace in these early years, it was the years following World War II that saw a tremendous expansion in generation-particularly in steam and hydro stations as illustrated by these statistics:

Table 1-1. Generation Capacity in the United States
(in millions of kW or gigawatts)

Year	Investor Utilities	Govn't Agencies	Non Utility	Total U.S.	Ave. gW Increase Per 5-yr. Period Ending ---
1950	55.2	13.7	6.9	82.9	4.0
1955	86.9	27.6	16.4	130.9	9.6
1960	128.5	39.6	17.8	185.8	11.0
1965	177.6	58.6	18.4	254.5	13.7
1970	262.7	78.4	19.2	360.3	21.2
1975	399.0	109.4	19.2	527.6	33.5
1980	477.1	136.6	17.3	631.0	20.7
1985	530.4	158.3	22.9	711.7	16.1
1990	568.8	166.3	45.1	780.2	13.7
1995	578.7	171.9	66.4	817.0	7.4
2000	443.9	192.3	23.2	868.2	10.3

NOTE: 1990-2000 divested generation from utilities to non-utility (merchant generation) probably affects allocation of generation

Net Summer Generating Capacity
(in millions of kW)

	Steam	Int. Comb.	Gas turb.	Nuclear	Hydro	Other	Total
1950	48.2	1.8	0	0	19.2	-	69.2
1960	128.3	2.6	0	0.4	35.8	-	167.1
1970	248.0	4.1	13.3	7.0	63.8	0.1	336.4
1980	396.6	5.2	42.5	51.8	81.7	0.9	578.6
1990	447.5	4.6	46.3	99.6	90.9	1.6	690.5
2000	507.9	5.9	83.8	97.6	99.1	17.4	811.7

Source: DOE statistics

PLANNING AND DEVELOPMENT

The period 1950-1990 was most important in the planning and development of electric generating stations. It began with the creation of many new stations and expansion of existing stations. Nuclear stations made their debut and subsequently at the end of the period were no longer acceptable in most areas of the United States. The oil crisis in the 70's had an effect on the use of oil fired units and created the need for intense electric conservation and alternative electric energy sources. Finally, the large central stations were being augmented by independent power producers and peaking units in smaller distributed area stations using waste heat from industrial processes, garbage fueled boilers, natural gas and methane gas from waste dumps.

The 1980's and 1990's also saw the effect of environmental restrictions and government regulation both on existing stations and new stations. Instead of a relatively short time to plan and build a new generating station, the process now takes 5-10 years just to secure the necessary permits—especially nuclear. Nuclear units grew from 18 stations totaling 7 million kW capacity to 111 units totaling almost 100 million kW. For now, it is not likely that many more new nuclear units will operate in the United States because of public opinion and the licensing process. The incident at Three Mile Island resulted in adverse public reaction despite the fact that safety measures built into the design and operation prevented any fatalities, injuries or environmental damage. The accident at Chernobyl added to the negative reaction despite the difference between the safer American design and the Russian nuclear design and operation.

Planning a new generating station in today's economic and regulatory climate is a very risky business because of the complicated and time consuming licensing process. Large capital investments are also being required to refit and modernize existing units for environmental compliance and to improve efficiencies. At the same time, more large sums of money are being spent on mandated conservation and load management (scheduling of consumer devices to achieve a lowered maximum demand) programs. These programs have affected the need for new generation or replacing older generation by significantly reducing the electrical energy requirements for system demand and total usage.

The future planning and development of electric generating stations will involve political, social, economic, technological and regulatory factors to be considered and integrated into an electrical energy supply plan. The system planner can no longer predict with the same degree of certainty when, where and how much generation capacity must be added or retired.

FUTURE CONSIDERATIONS

Will new transmission capacity be added and coordinated with generation changes since the declining trend of generation additions has followed the trend of generation additions? What will be the impact of large independent transmission regional operators on system reliability?

With new generation added principally in the form of relatively high cost per kWh combustion turbines and not lower cost base load steam. turbine units or hydro, will deregulation result in lower unit energy costs to customers?

Can reduction in system load through conservation measures be forecast accurately and timely enough to allow for adequate generation? Can conservation reliably replace generation?

Will the merchant generators and energy companies contribute towards research programs aimed at improving reliability and reducing costs? Previous utility active support of the Electric Power Research Institute with money and manpower resulted in many industry advances in the state of the art, but will this continue?

PRESENT POWER PLANT CONSIDERATIONS

Many factors, all interrelated, must be considered before definite plans for a power plant can be made. Obviously the final construction will contain a number of compromises each of which may influence the total cost but all are aimed at producing electrical energy at the lowest possible cost. Some factors are limited as to their variation such as available sites. Plans for the expansion of existing stations also face similar problems although the number of compromises may be fewer in number.

SITE SELECTION

For minimum delivery losses a plant site should be close as possible to the load to be served as well as minimizing the associated expensive transmission costs connecting the plant to the system. Environmental restrictions and other possible effects on overhead electric lines are requiring more underground connections at a significantly higher cost. Site selection must also include study of future expansion possibilities, local construction costs, property taxes, noise abatement, soil characteristics, cooling water and boiler water, fuel transportation, air quality restrictions and fuel storage space. For a nuclear station additional factors need to be considered: earthquake susceptibility, an evacuation area and an emergency evacuation plan for the surrounding community, storage and disposal of spent fuel, off-site electrical power supply as well as internal emergency power units and most important the political and community reception of a nuclear facility. If a hydro plant is to be considered, water supply is obviously the most important factor. Compromise may be required between the available head (height of the available water over the turbine) and what the site can supply. As in fossil fuel and nuclear plants the political and public reception is critical.

After exhaustive study of all these factors the first cost is estimated as well as the annual carrying charges which include the cost of capital, return on investment, taxes, maintenance, etc. before the selection decision can proceed.

SELECTION OF POWER STATION UNITS

The first selection in a new unit would be the choice between a base load unit or a peaking unit. Most steam stations are base load units—that is they are on line at full capacity or near full capacity almost all of the time. Steam stations, particularly nuclear units, are not easily nor quickly adjusted for varying large amounts of load because of their characteristics of operation. Peaking units are used to make up capacity at maximum load periods and in emergency situations because they are easily brought on line or off line. This type of unit is usually much lower in first cost than a base load unit but is much higher in energy output cost. Peaking units are most likely to be gas turbines, hydro or internal combustion units. Reciprocating steam engines and internal combustion powered plants are generally used for relatively small power stations because of space requirements and cost. They are sometimes used in large power stations for starting up the larger units in emergencies or if no outside power is available. Nuclear power stations are mandated to have such emergency power sources. No further discussion of this type of unit will be made.

Steam Power Plants

Steam power plants generally are the most economical choice for large capacity plants. The selections of boilers for steam units depends greatly on the type of fuel to be used. Investment costs as well as maintenance and operating costs which include transportation and storage of raw fuel and the disposal of waste products in the energy conversion process. Selection also depends on the desirability of unit construction-one boiler, one turbine, one generator-or several boilers feeding into one common steam header supplying one or more turbine generators. Modern plant trends are towards the unit type construction. For nuclear plants the cost of raw fuel, storage and disposal of spent fuel is a very significant part of the economics.

Hydro-electric Plants

With some exceptions, water supply to hydro plants is seasonal. The availability of water may determine the number and size of the units contained in the plant. Unless considerable storage is available by lake or dam containment, the capacity of a hydro plant is usually limited to the potential of the minimum flow of water available. In some cases hydro plants are designed to operate only a part of the time. Other large installations such as the Niagara Project in New York operate continuously. In evaluating the economics of hydro plants the first cost and operating costs must also include such items as dam construction, flood control, and recreation facilities.

In times of maximum water availability, hydro plants may carry the base load of a system to save fuel costs while steam units are used to carry peak load variations. In times of low water availability the reverse may prove more economical. The difference in operating costs must be considered in estimating the overall system cost as well as system reliability for comparison purposes.

CONSTRUCTION COSTS

Construction costs vary, not only with time, but with locality, availability of skilled labor, equipment, and type of construction required. For example in less populated or remote areas skilled labor may have to be imported at a premium; transportation difficulties may bar the use of more sophisticated equipment; and certain parts of nuclear and hydro plants may call for much higher than normal specifications and greater amounts than is found in fossil plants. Seasonal variations in weather play an important part in determining the costs of construction. Overtime, work stoppages, changes in codes or regulations, "extras" often appreciably increase costs but sometimes unforeseen conditions or events make them necessary. Experience with previous construction can often anticipate such factors in estimating costs and comparing economics.

FUEL COSTS

Since the cost of fuel is often one of the larger parts of the overall cost of the product to the consumer, it is one of the basic factors that determine the kind, cost and often the site of the generating plant. The cost attributed to the fuel must also include its handling/transportation/storage charges and should as much as practical take into account future fluctuations in price, continued availability and environmental restrictions. For instance the oil crisis in the 70's sharply escalated the cost of fuel for oil fired plants and limited its supply. In the years following further cost escalations resulted from the environmental requirements for lower sulfur fuels.

FINANCE COSTS

Like other items in the construction, maintenance and operation of a power plant, the money to pay for them is obtained at a cost. This includes sale of bonds and stocks, loans and at times the reinvestment of part of the profits from operations of the company. Even if the entire cost of the proposed plant was available in cash, its possible earning potential invested in other enterprises must be compared to the cost of obtaining funds by other means such as those mentioned previously before a decision is made on how to finance the project.

In this regard, the availability of money at a suitable cost often determines the schedule of construction. This may occur from the absolute lack of capital or because of exorbitant interest rates. In some cases the total cost for obtaining the required funds may be lower if it is obtained in smaller amounts over a relatively long period of time. In an era of inflation, the reverse may be true and the entire amount obtained at one time and accelerating construction to reduce the effects not only of the cost of money but increasing costs of labor and material. In this regard it may be worth knowing what a dollar today at a certain interest rate is worth X years hence. Conversely, what a dollar invested X years hence is worth today at a certain interest rate. These are given in the following formulas:

$$\text{Future worth} \;=\; (1 + \text{interest rate})^x$$

$$\text{Present worth} \;=\; \frac{1}{(1 + \text{interest rate})^x}$$

The impact of taxes, federal, state, and local, and others (income, franchise, sales, etc.) and insurance rates may also affect the method of financing and construction.

FIXED COSTS

It should be of interest to note that there are certain costs associated with a generating plant that must be provided for whether or not the plant produces a single kilowatt-hour of electrical energy. This is similar to the expenses of car owners who must meet certain expenses even though the car may never be driven.

Fixed charges are those necessary to replace the equipment when it is worn out or made obsolete. Interest and taxes carry the investment, while insurance and accumulated depreciation funds cover the retirement of physical property.

Interest is the time cost of money required for the work. It is affected by the credit rating of the utility, the availability of money and other financial conditions internal and external to the company at the time money is borrowed.

Taxes can be as variable as interest rates. In addition to property taxes, utilities are also subject to franchise tax, income tax, licensing and other special imposts created by local, state and federal authorities.

Insurance carried by utilities include accidents, fire, storms, vehicles and particularly in the case of power plants boiler insurance. Insurance carriers for boilers require periodic inspections by their personnel and may result in recommendations for changes or improvements to the plant the benefits of which can offset the cost of insurance.

Depreciation of property and equipment takes place continually. At some time after initial installation most equipment will reach a condition at which it has little or no useful life remaining. A retirement reserve permits replacement to retain the integrity of the initial investment.

Chapter 2

Electric Power Generation

ENERGY CONVERSION

Power generating plants, like other manufacturing plants, process raw materials into useful products, often accompanied by some waste products. For power plants the useful product is electrical energy. The waste products for fossil plants include ash and smoke visibly and heat invisibly.

Similarities include the use of equipment and materials that serve to expedite and improve efficiency of operations, although they may not be directly involved in the manufacture of the product. For example, water may be used to produce the steam and for cooling purposes, oil to lubricate moving parts, and fans and pumps to move gases and fluids. Additional similarities include facilities for the reception of raw materials, disposal of waste, and for delivery of the finished product as well as trained personnel to operate the plant. Economic considerations, including capital investment and operating expenses which determine the unit costs of the product while meeting competition (oil, natural gas in this case) are common to most business enterprises.

There are some important dissimilarities. As a product electricity not only is invisible and hazardous in its handling but for the most part cannot be stored. Inventories cannot be accumulated and the ever changing customer demands must be met instantly. All of this imposes greater standards of reliability in furnishing a continuing supply both in quantity and quality. This criteria assumes even greater importance as such generating plants are vital to national economy and contribute greatly to the standard of living.

In the larger central generating plants, fossil or nuclear energy (in the form of fuel) is first converted into heat energy (in the form of steam), then into mechanical energy (in an engine or turbine), and finally into electrical energy (in a generator) to be utilized by consumers. A schematic arrangement is shown in Figure 2-1, below.

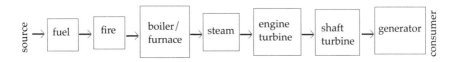

Figure 2-1
Schematic Diagram of Energy Conversion

Most commonly, electricity is produced by burning a fossil fuel (coal, oil or natural gas) in the furnace of a steam boiler. Steam from the boiler drives a steam engine or turbine connected by a drive shaft to an electrical generator.

A nuclear power plant is a steam-electric plant in which a nuclear reactor takes the place of a furnace and the heat comes from the reaction within the nuclear fuel (called fission) rather than from the burning of fossil fuel. The equipment used to convert heat to power is essentially the same an ordinary steam-electric plant. The product, electrical energy is identical; see Figure 2-2.

The processes and the equipment to achieve these energy transformations will be described in fundamental terms, encompassing arrangements and modifications to meet specific conditions. Some may be recognized as belonging to older practices (for example burning lump coal on iron grates). While serving purposes of illustration, it must be borne in mind that for a variety of reasons, some of the equipment and procedures continue in service and, hence, knowledge of their operation is still desirable. Pertinent changes, developments and improvements, brought about by technological, economic and social considerations are included.

The four conversion processes in a typical steam generating plant may be conveniently separated into two physical entities, following accepted general practice. The first two processes comprise operations known as the BOILER ROOM, while the latter two are included in those known as the TURBINE ROOM.

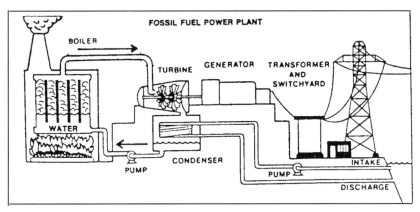

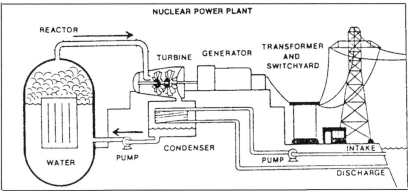

Except for the source of heat they use to create steam, nuclear and fossil power plants are basically the same.

Figure 2-2. (*Courtesy LI Lighting Co.*)

INPUT ENERGY SOURCES

Sources of energy for the production of electricity are many and varied. In addition to the energy contained in falling water, the more common are contained in fuels which contain chemical energy. These can be characterized as fossil and non-fossil fuels; the former, formed from animal and plant matter over thousands of years, while the latter comprises radioactive-associated materials. Coal, oil and natural gas fall into the first category as fossil fuels, while uranium and plutonium (and less known thorium) comprise so-called nuclear fuels. All fuels may be classified as solid, liquid or gaseous, for handling purposes.

A number of fuels commonly employed in the production of

electricity are contained in Table 2-1; representative values of their heat content and the components of the chemical compounds are indicated.

Table 2-1. Energy Sources

1. *Hydro* - Depends on availability, volume and head (distance from the intake to the water wheels).

2. *Fossil Fuels* - Typical Characteristics (Approximate Values)

   Components in Percent

Fuel	Btu/lb	Carbon	Hydrogen	Oxygen	Sulphur	Nitrogen	Ash
Wood	9,000	52.0	6.0	25.0	0.3	15.0	1.7
Coal Lignite	11,000	60.0	6.0	4.0	1.0	1.0	28.0
Oil	18,500	88.0	8.2	0.5	3.0	0.1	0.2
Natural Gas	22,000	69.0	23.5	1.5	0.3	5.7	0.0

3. *Nuclear Fuels*
   Fuel Btu/lb
   Uranium
   Plutonium   } 38.7 billion
   Thorium

4. *Other Fuels*
   Fuel
   Tar
   Garbage
   Manure Heap Gas
   Aquatic & Land Plants
   Bagasse
   Husks

   Components may vary widely, but the combustible contents generally include carbon and hydrogen as basic elements.

5. *Other Energy Sources*
   Geo-thermal
   Wind
   Tides
   Solar (Direct Sun Rays)
   Temperature Differences
   between Surface & Deep
   Layers of Water Bodies

   Variable - from very large quantities to very small quantities - Some not always available.

COMBUSTION

Combustion, commonly referred to as burning, is the chemical process that unites the combustible content of the fuel with oxygen in the air at a rapid rate. The process converts the chemical energy of the fuel into heat energy, and leaves visible waste products of combustion, generally in the form of ash and smoke.

Chemistry of Combustion

In order to understand what takes place when fuel is burned, it is desirable (though not essential) to review the chemistry of the action involved.

Elements, Molecules and Atoms

All substances are made up of one or more "elements." An element is a basic substance which, in present definition, cannot be subdivided into simpler forms. The way these elements combine is called chemistry.

The smallest quantity of an element, or of a compound of two or more elements, is considered to be the physical unit of matter and is called a "molecule." In turn, molecules are composed of atoms. An atom is defined as the smallest unit of an element which may be added to or be taken away from a molecule. Atoms may exist singly but are usually combined with one or more atoms to form a molecule. Molecules of gaseous elements, such as oxygen, hydrogen and nitrogen, each consist of two atoms. See Figure 2-3.

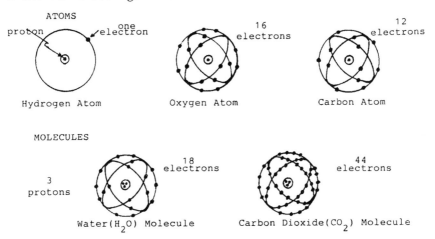

Figure 2-3. Illustrating Composition of Atoms & Molecules

Chemical Combinations

Chemical symbols and equations convey much information in a condensed form. Chemical combinations may be written as equations in which, for convenience, symbols are used to represent different elements. Thus, for the gaseous elements mentioned above: O represents one atom of oxygen, H one atom of hydrogen, and N one atom of nitrogen. As indicated above, molecules of gaseous elements each contain two atoms and are represented by O_2; H_2; and N_2. Molecules of non-gaseous elements may consist of a single atom; for example, carbon represented by C, sulphur by S, etc.

Water is represented by H_2O which indicates that each molecule of which it is composed consists of two atoms of hydrogen and one of oxygen. Two molecules of water would be indicated by placing the number 2 in front of the symbol - thus, $2H_2O$.

Atoms of the different elements have different relative "weights;" hence chemical combinations always take place in definite proportions. For example, hydrogen combines with oxygen to form water: these two elements combine in the proportion of two atoms of hydrogen to one atom of oxygen and it will be found that 2 pounds of hydrogen will combine with 16 pounds of oxygen to make 18 pounds of water. If more hydrogen is present, it will remain uncombined; if less, some of the oxygen will remain uncombined. The parts entering into combination will always be in the proportion of one to eight.

Atomic and Molecular Weights

Since the elements combine in certain definite proportions, a value has been assigned to each one to simplify the computations. This value is called the "atomic weight." Hydrogen, being the lightest known element, has been taken as one or unity, and the heavier elements given weights in proportion. The following atomic weights are used in fuel combustion problems:

Hydrogen (H) 1
Carbon (C) 12
Nitrogen (N) 14
Oxygen (O) 16
Sulphur (S) 32

The molecular weight of a substance is obtained by the addition of the atomic weights composing the substance. Thus, carbon dioxide, CO_2, has a molecular weight of one carbon atom and two oxygen atoms:

$$1 \times 12 + 2 \times 16 = 12 + 32 = 44$$

that is, the molecular weight of carbon dioxide is 44.

Fuel and Air

Fossil fuels are composed of carbon (C), hydrogen (H), sulphur (S), nitrogen (N), oxygen (O), and some other important elements. As indicated previously, the largest percentage of such fuels is pure carbon and the next largest part is hydrocarbons composed of hydrogen and carbon in varying proportions, depending on the kind of fuel.

Air is mainly a combination of two elements, oxygen (O) and nitrogen (N), existing separately physically and not in chemical combination. Neglecting minor quantities of other gases, oxygen forms 23.15 percent of air by weight, nitrogen forming the other 76.85 percent. On the other hand, the volumes of the two gases would be in proportion of 20.89 percent oxygen and 79.11 percent nitrogen; or a cubic foot of air would contain 0.2089 cubic foot of oxygen and 0.7911 cubic foot of nitrogen.

Combustion may be defined as the rapid chemical combination of an element, or group of elements, with oxygen. The carbon, hydrogen and sulphur in the fuel combines with the oxygen in the air and the chemical action can give off a large quantity of light and heat. The carbon, hydrogen and sulphur are termed combustibles. If the air was pure oxygen and not mixed with the inert nitrogen gas, combustion once started, would become explosive as pure oxygen unites violently with most substances - with damaging effects to boilers and other combustion chambers.

Chemical Reactions—Combustion Equations

The principal chemical reactions of the combustion of fossil fuels are shown in the following equations expressed in symbols:

(1) Carbon to carbon monoxide $\quad\quad 2C + O_2 = 2CO$
(2) Carbon to carbon dioxide, $\quad\quad\quad 2C + 2O_2 = 2CO_2$
(3) Carbon monoxide to carbon dioxide $\quad 2CO + O_2 = 2CO_2$
(4) Hydrogen to water $\quad\quad\quad\quad\quad 2H_2 + O_2 = 2H_2O$
(5) Sulphur to sulphur dioxide $\quad\quad\quad S + O_2 = SO_2$

(1) For incomplete combustion of C to CO:

	2C	+	O_2	=	2CO	N_2	Air	Total Combustion Products
Atom Wt	24		32		56	—	—	—
Pounds	1		1.333		2.333	4.425	5.758	6.758

(2) For complete combustion of C to CO_2:

	C	+	O_2	=	CO_2			
Atom Wt	12		32		36	—	—	—
Pounds	1		2.667		3.667	8.854	11.521	12.521

(3) For combustion of CO to CO_2

	2CO	+	O_2	=	$2CO_2$			
Atom Wt	56		32		88	—	—	—
Pounds	1		0.571		1.571	1.892	2.463	3.463

(4) For combustion of H_2 to H_2O:

	$2H_2$	+	O	+	$2H_2O$			
Atom Wt	4		32		34	—	—	—
Pounds	1		8.000		9.000	26.557	34.557	35.557

(5) For combustion of S to SO_2:

	S	+	O_2	=	SO_2			
Atom Wt	32		32		64	—	—	—
Pounds	1		1.000		2.000	3.320	4.320	5.320

From these unit figures, the air requirements and the amount of combustion products (flue gases) may be determined for the different kinds of fuels.

Other Chemical Reactions

The chemical reactions indicated above may be influenced by the relative amounts of the elements present.

Thus, if a small amount of carbon is burned in a great deal of air, CO_2 results. But if there is a great deal of carbon and a little air, instead of a small portion of the carbon being burned to CO_2 as before and the rest remain unaffected, a large portion will be burned to CO and there will be no CO_2 formed at all. The action is as if CO was a nearly satisfied combustion, and if an unattached O atom should be present and nothing else available, it would unite with the CO and form CO_2. But if more C was present, the C atom would prefer to pair off with an atom of C and make more CO. Such preferences are referred to as "affinity."

Now if CO_2 comes into contact with hot carbon, an O atom, acting in accordance with the affinity just mentioned, will even leave the already existing CO_2 and join with an atom of hot carbon, increasing the amount of CO in place of the former CO_2 molecule.

Such action may actually take place in parts of the fuel being burned and affect the design and operation of furnaces.

Perfect Combustion vs Complete Combustion

Perfect combustion is the result of supplying just the right amount of oxygen to unite with all the combustible constituents of the fuel, and utilizing in the combustion all of the oxygen so supplied that neither the fuel nor the oxygen may be left over.

Complete combustion, on the other hand, results from the complete oxidation of all the combustible constituents of the fuel, without necessarily using all the oxygen that is left over. Obviously, if extra oxygen is supplied, it must be heated and will finally leave the boiler carrying away at least part of the heat, which is thereby lost. If perfect combustion could be obtained in a boiler there would be no such waste or loss of heat. The more nearly complete combustion can approach a perfect combustion, the loss will occur in the burning of a fuel. The problems of design and operation of a boiler are contained in obtaining as nearly as possible perfect combustion.

HEAT AND TEMPERATURE

When fuels are burned, they not only produce the combustion products indicated in the chemical equations listed above. More importantly, they also produce heat. The heat will cause the temperature of the gases and the surrounding parts to rise.

The distinction between temperature and heat must be clearly understood. Temperature defines the *intensity*, that is, how hot a substance is, without regard to the *amount* of heat that substance may contain. For example, some of the boiling water from a kettle may be poured into a cup; the temperature of the water in the kettle and the cup may be the same, but the amount of heat in the greater volume of water in the kettle is obviously several times the amount of heat contained in the water in the cup.

If two bodies are at different temperatures, heat will tend to flow from the hotter one to the colder one, just as a fluid such as water tends to flow from a higher to a lower level.

Temperature may be measured by its effect in expanding and contracting some material, and is usually measured in degrees. The mercury thermometer is a familiar instrument in which a column of mercury is enclosed in a sealed glass tube and its expansion and contraction measured on an accompanying scale. Two such scales are in common use, the Fahrenheit (F) and the Centigrade or Celsius (C). The former has the number 32 at the freezing point of water and 212 at the boiling point; thus 180 divisions, or degrees, separate the freezing and boiling points or temperatures of water. The latter has the number zero (0) at the freezing point of water and 100 at the boiling point; thus 100 at the boiling points thus 100 divisions or degrees separate the freezing and boiling points or temperatures of water. Both scales may be extended above the boiling points and below the freezing points of water. Refer to Figure 2-4. Other instruments may employ other liquids, gases, or metals, registering their expansion and contraction in degrees similar to those for mercury. Temperature values on one scale may be converted to values on the other by the following formuas:

$$°C = \frac{5}{9}\left(°F - 32\right) \text{ and } °F = \frac{9}{5}\left(°C + 32\right)$$

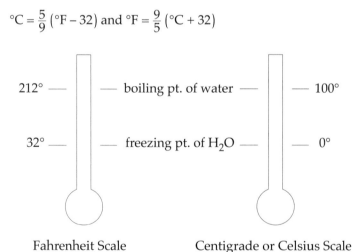

Fahrenheit Scale   Centigrade or Celsius Scale

Figure 2-4. Comparison of F & C Temperature Scales

Heat, that is, the amount of heat in a substance, may be measured by its effect in producing changes in temperature of the substance. Thus, if heat is added to water, it will become hot and its temperature rise. The unit of heat, or the amount of heat, is measured in British Thermal Units (Btu) or in Calories (C). The Btu is the amount of heat required to raise one pound of water (about a pint) one degree Fahrenheit. The Calorie is the amount of heat required to raise one kilogram of water (about 1-1/3 liters) one degree Centigrade. Heat values in one system may be converted to values in the other by the following formulas:

1 Btu = 0.2521 Calorie   and   1 Calorie = 3.9673 Btu's
1 Btu = Approx. 1/4 Calorie   and   1 Calorie = Approx. 4 Btu's

Different substances require different amounts of heat to raise the temperature one degree; these quantities are called the specific heats of the substances. Compared to water (its specific heat taken as 1), that of iron is 0.13, kerosene 0.5, air 0.244, etc.

Temperature of Ignition

When air is supplied to a fuel, the temperature must be high enough or ignition will not take place and burning will not be sustained. Nothing will burn until it is in a gaseous state. For example, the wax of a candle cannot be ignited directly; the wick, heated by the flame of a match, draws up a little of the melted wax by capillary action until it can be vaporized and ignited. Fuels that liquefy on heating usually will melt at a temperature below that at which they ignite. Solid fuels must be heated to a temperature at which the top layers will gasify before they will burn.

Table 2-2 gives ignition temperatures for various substances, including some hydrocarbons mentioned earlier.

The ignition temperatures of fuels depend on their compositions; since the greatest part is carbon, the temperature given in the table, 870°F, will not be far wrong. Heat must be given to the fuel to raise it to the temperature of combustion. If there is moisture in the fuel, more heat must be supplied before it will ignite, since practically all of the moisture must be evaporated and driven out before the fuel will burn.

Temperature of Combustion

When the fuel is well ignited, its temperature will be far above

Table 2-2. Ignition Temperatures

Substance	Symbol	Approximate Ignition Temperature	
		°F	°C
Sulphur	S	470	245
Carbon (fixed)	C	870	465
Acetylene	C_2H_2	900	480
Ethylene	C_2H_4	950	510
Ethane	C_2H_6	1000	540
Hydrogen	H_2	1130	610
Methane	CH_4	1200	650
Carbon Monoxide	CO	1210	655

that of ignition. While combustion is taking place, if the temperature of the elements is lowered (by whatever means) below that of ignition, combustion will become imperfect or will cease, causing waste of fuel and the production of a large amount of soot.

Since it is the purpose to develop into heat all the latent energy in the fuel, it is important that the temperature of the fuel be kept as high as practical. The maximum temperature attainable will depend generally on four factors:

1. It is impossible to achieve complete combustion without an excess amount of air over the theoretical amount of air required, and the temperature tends to decrease with the increase in the amount of excess air supplied.

2. If this excess air be reduced to too low a point, incomplete combustion results and the full amount of heat in the fuel will not be liberated.

3. With high rates of combustion, so much heat can be generated that, in a relatively small space, even if the excess air is reduced to the lowest possible point, the temperatures reached may damage the containing vessel.

4. Contrary-wise, if the containing vessel is cooled too rapidly (by
    whatever means), the temperature of the burning fuel may be low-
    ered resulting in poor efficiency.

The rate of combustion, therefore, affects the temperature of the
fire, the temperature increasing as the combustion rate increases, provided that relation of fuel to air is maintained constant.

Temperatures of 3000°F may be reached at high rates of combustion and low amounts of excess air and may cause severe damage to
heat resisting materials and other parts of the containing vessel.

COMBUSTION OF FOSSIL FUELS

Carbon and hydrogen are the only elementary fuels; sulphur and
traces of other elements do burn and give off heat, but these are col-
lectively so small as to be considered negligible and their constituents as
nuisance impurities. Oxygen from the air is the only elementary burning
agent necessary to every heat producing reaction.

Coal

Coal is a complex substance containing mainly carbon, hydrogen,
sulphur, oxygen and nitrogen. Typical approximate percentages of these
constituents are shown in Table 2-1. It must be recognized, however,
that there are different kinds of coals whose characteristics may differ
from the typical values shown. Coals are classified with the aid of the
various characteristics by which they may be distinguished. Among the
more important are:

Volatile Matter - This term describes the mixture of gases and hydrocarbon vapors that may be given off when coal is heated at very high
temperatures; these may include acetylene, ethylene, ethane, methane,
and others. The more the volatile matter, the more liable the coal will
produce smoke. This is an indicator of the property of the coal.

Fixed Carbon - This term applies to that portion of the carbon left
in the coal after the volatile matter is subtracted. It is usually combined
with the percentages of moisture and ash in the coal when classifying
coals.

Coals may have different properties depending on where they are
mined and require different ways of firing for best results. In addition to

heating values, the amounts of ash and smoke produced when the coal is fired are different for the several kinds of coal. Their classification is indicated in Table 2-3.

Ash - Ash is the residue after all the carbon, hydrogen and the small quantities of sulphur are burned away; coal ash varies considerably in the different kinds of coal. A large amount of ash in coal not only lowers its heating value but causes ash removal to become more difficult. The clinkering of coal depends largely on its composition. Ash may consist of alumina and silica derived from clay, shale and slate, iron oxide from pyrites or iron sulfide, and small quantities of lime and magnesia.

Heat of Combustion

If one pound of carbon is completely burned to CO_2, the heat developed will be 14.550 Btu. That is, the heat evolved would heat 14,550 lbs of water 1°F, or 1455 lbs 10°F, or 145.5 lbs 100°F, etc. The heat of combustion of carbon is therefore said to be 14.550 Btu. (See equation 2-2, p. 18.)

If, however, the pound of carbon is incompletely burned to CO_2, which occurs when too little air is supplied, the heat involved would be only 4,500 Btu. If the gas passes off, the loss will be 14,550 minus 4,500 or 10,050 Btu, or approximately 70% of the possible heat that can be evolved. That is, 70% of the fuel value is wasted if CO is formed instead of CO_2. (See equation 1, p. 18)

The CO, however, can be burned to CO_2 by the addition of more air under suitable temperature conditions and, in the end, the same amount of heat per pound of carbon, 14,550 Btu, will be developed as

Table 2-3. Classification of Coals

Kind of Coal	Volatile Matter in %	Fixed Carbon, Ash & Moisture Content-%	Heating Values Approx. Btu
Hard Anthracite	3 to 7.5	97.0 to 92.5	11500-13500
Semi Anthracite	7.5 to 12.5	92.5 to 87.5	13300-13800
Semi Bituminous	12.5 to 25.0	87.5 to 75.0	14500-15200
Bituminous-East	25 to 40	75 to 60	14000-14500
Bituminous-West	35 to 50	65 to 50	13000-14000
Lignite	Over 50	Under 50	10500-13000

when C is burned to CO_2 in one operation. (See equation 3, p. 18)

The heat of combination of hydrogen is 62,100 Btu per pound and of sulphur 4,050 Btu per pound.

The heat of combustion of a typical sample of dry coal is given below in Table 2-4, analyzed by its components; the contributions of each to the total heat evolved by a pound of coal is also indicated.

Table 2-4. Ultimate Analysis of Coal Sample

Substance	% by weight	Heat of Combustion Btu/lb	Heat per lb of coal-Btu
Carbon (C)	84.0	14,550	12,222
Hydrogen (H)	4.1 - 0.3*	62,100	2,360*
Sulphur (S)	0.8	4,050	32
Oxygen (O)	2.4*	-	-
Nitrogen (N)	1.4	-	-
Ash (Inert)	7.3	-	-
Total	100.0		14,614

*Hydrogen in the fuel is 4.1%, but some of which is combined with oxygen and exists as inherent moisture which is not available to produce heat. Since it takes 8 parts of oxygen to convert 1 part of hydrogen to water, then 2.4 divided by 8 = 0.3 parts of hydrogen will not produce heat; 4.1 - 0.3 = 3.8% of hydrogen will produce heat.

If instead of being dry, the coal is burned as it comes, a condition termed "as fired, " it will have some moisture in it. Since heat is required to evaporate the moisture, and the heat thus used is not available, the effective Btu per pound of coal will be less. For example, if the coal contained 5% moisture, the heating value would be 14,614 × 0.95 or 13883 Btu. (Actually, with the added moisture, the percentages of the several constituents would also be slightly altered to keep the total to 100 percent.)

When coal is burned, the temperature of the burning fuel may be judged roughly from the color, brilliancy and appearance of the fire. For example:

Dull red to cherry red ....1000 to 1500°F
Bright cherry red.............................. 1500 to 1600 °F
Orange...1600 to 1700°F
Light orange to yellow ....1700 to 2000°F
White to dazzling white........................Over 2000°F

GASIFICATION OF COAL

The gasification of coal is a process that converts solid coal into combustible gas mainly composed of carbon monoxide and hydrogen. The gas, cleaned of waste material and particulate matter (pollutants), is ready to be burned to produce heat and energy.

Coal, together with hydrogen and steam, heated to a temperature of some 1600°C will produce a gas of some 20 to 25 percent Btu content of natural gas (depending on the type of coal), together with some important by-products. The gas, a mixture of carbon monoxide and hydrogen, synthesis gas or syngas, that is ultimately burned in a steam boiler to produce steam to power one or more steam turbines to generate electricity. (Refer to Figure 2-5)

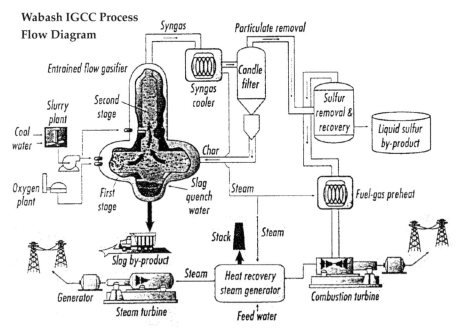

Figure 2-5. Variations of Gasification Process (Courtesy *Power Engineering*)

Piñon Pine's IGCC Process Flow Diagram

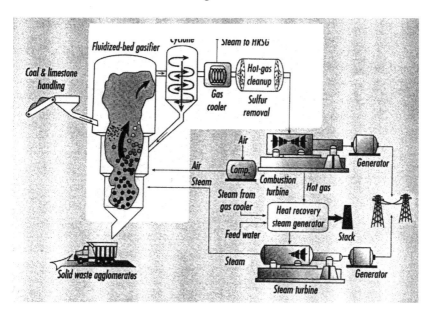

Process Flow of the Polk IGCC Project

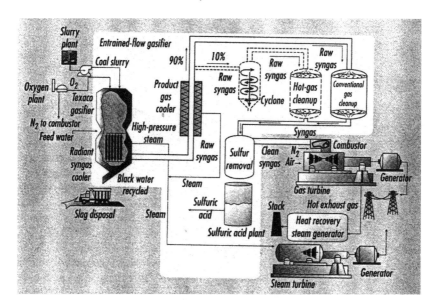

Figure 2-5. (*Continued*)

Although of low Btu, this gas can compete economically with other fuels because:

1. The by-products of the process beside particulates (coal dust) are sulfur and nitrogen compounds that are extricated at this point (and other residues and the gas itself are valuable in the petrochemical industry). These compounds are pollutants that are also produced in the combustion of other fuels, requiring expensive processes to eliminate or reduce them to values meeting environmental require-ments before released to the atmosphere. Syngas when burned produces carbon dioxide and water, requiring no processing when released to the atmosphere.

2. The syngas, produced at high pressure, is made to power a gas turbine to generate electricity that constitutes a portion of the over-all electricity output. It is then exhausted at a lower pressure, goes on to be burned in steam boilers to produce steam to drive other steam turbines to produce electricity. This is sometimes referred to as a composite or combined cycle. (Refer to Figure 2-6)

3. The syngas produced, not only fires the gasifier unit itself, but, as mentioned, produces valuable by-products that include sulfur and nitrogen oxides and other materials. Sulfur nitrogen and other mate-rials are recovered by standard processes that may employ cyclone and ceramic filters, catalysts, scrubbers, absorption and dissolution in amino acids and other solutions. An inert slag is also produced that may be used as an aggregate in highway and other construction.

In the use of coal as a fuel, a comparison of the gasified method to that of coal burned using wet slurry feed, the typical components of raw syngas and slurry coal are shown in the Table 2-5.

Some 80% of the energy in the original coal is contained in the syngas, as it leaves the gasification process, and is known as the cold gas efficiency. Another 15% from the steam produced in the gasifier and gasifier cooler, for a total of some 95% in comparison with a cold gas efficiency of some 75% using wet slurry feed. In the gasifier, the syngas is cooled to some 800-900°C by gas flowing through the unit, and then further cooler to some 400°C in a high pressure heat exchanger (the waste heat delivered to a boiler producing steam to power a steam tur-

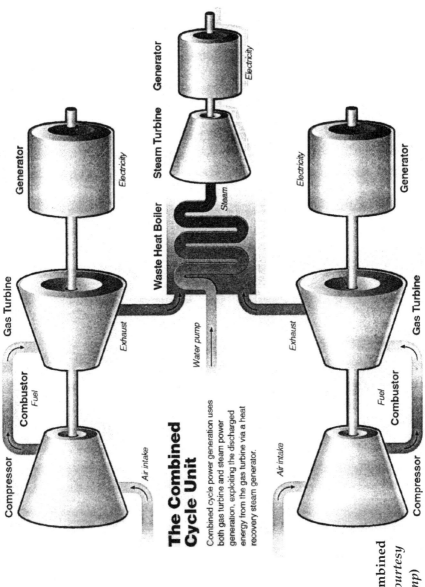

The Combined Cycle Unit

Combined cycle power generation uses both gas turbine and steam power generation, exploiting the discharged energy from the gas turbine via a heat recovery steam generator.

Figure 2-6. Combined Cycle Unit (*Courtesy Exxon The Lamp*)

Table 2-5. Comparison of Syngas and Coal Components (% volume)

Material	Syngas	Slurry Coal
Hydrogen H_2	26.7	30.3
Carbon Monoxide CO	63.3	39.7
Carbon Dioxide CO_2	1.5	0.8
Methane CH_4	0,0	01
Hydrogen Sulfide H_2S	1.3	1.0
Nitrogen N_2	4.1	0.7
Argon Ar	1.1	0.9
Water H_2O	2.0	16.5

Source: Royal Dutch Shell Group

bine to generate electricity). The syngas may then be routed through an intermediate pressure heat exchanger. This produces steam for another steam turbine, lowering the syngas temperature to some 250°C. The raw syngas at 250°C is then passes through filters to remove coal particulates and fly ash, which are returned to the gasifier unit, resulting in almost 100% carbon conversion.

The syngas is finally burned in a steam boiler with "clean" emission the result of combustion. Various processes remove the sulfur, nitrogen, and other components of the coal before the gas is burned in the steam boiler. The gasification of coal not only significantly reduces the pollution normally associated with generation of electricity from coal, but also opens up the vast coal reserves to active economic competition with other energy sources.

Earlier gasification systems were proven impractical, mainly because pollutants were extricated externally to the operation and the gas produced waste pressures insufficient for direct turbine operation. Also, the gasification vessel was not self-cleaning because of the relatively low temperatures involved, and the system had to be shut down for routine cleaning and heat brought up to speed on restarting. The principal such systems were the Lurgi (original), Winkler, and the Koppers-Totzen systems. They all employed steam and oxygen (air) acting on coal at different temperatures and pressures, providing various combinations of carbon monoxide, carbon dioxide, methane and hydrogen. The Lurgi system, because of the relatively low pressures (30-35 ATM) and temperatures (980°C) was most suitable for Lignite and Bituminous coal, because the steam and low temperature prevented the melting of ash. The Winkler system was essentially the same as the Lurgi, but operated on one ATM pressure. The Koppers-Totzen system operated at temperatures of 1600-1900°C with the time the acting on the coal (because of smaller particles) at about one second; the high temperature increased the reaction time allowed any type coal to be used in the system. Other relatively minor differences included the methods of introducing the coal, steam and oxygen into the gasifier vessel.

FUEL OIL

Fuel oil is the principal liquid fuel used for the production of electric power, although some residual tars from some industrial processes are sometimes also used.

Fuel oils are the heavy hydrocarbons remaining after the lighter hydrocarbons (such as gasoline, naptha and kerosene) have been removed from the crude oil. The composition of the fuel oil varies greatly, depending on the origin of the crude oil and on the degree to which the distillation has been carried.

Classification

Fuel oils may be divided into three classes on the basis of the principal chemical constituents which remain after distillation.

1.   Paraffin base—mainly found in eastern United States fields.

2.   Mixed paraffin and asphalt—midwest and southwest United States and some mid-east fields.

3.   Asphalt base—mainly found in Gulf of Mexico fields in Texas, Mexico, northers South America, and in some mid-east, far-east (Malaysian), and Siberian fields.

Paraffin base oil is generally more expensive and uneconomic for use as a fuel as it yields an abundance of valuable lighter lubricating oils. Mixed base oils, because of their asphalt content, are more costly to refine as highly as pure paraffin oil, and are often used as fuel for heating and for heavy-oil engines. Asphalt base oils constitute practi-

cally all of the fuel used for steam and power production purposes.
Fuels are further classified according to their degree of refinement:

1. Distillates   ⎫   These are rarely used for electrical power
2. Topped Crudes   ⎬   generation because of their high cost.
3. Residual Oils   ⎭   Sometimes also known as Bunker C oils, gener-
                            ally used in power plants.
4. Blended Oils   Often resorted to in order to reduce the high
                            viscosity of the residual oils.

Properties of Fuel Oils
        Certain characteristics of fuel oils and the manner in which they
affect suitability for use in power generation may be summarized as
follows:

Specific Gravity
        This is simply the weight of oil compared to the weight of an equal
volume of water, both at 60°F. It is commonly measured in degrees API
(American Petroleum Institute) and is measured by a hydrometer-like
instrument. As oil is frequently sold on the basis of volume delivered,
corrected to 60°F, its specific gravity must be known to make proper
correction.

Viscosity
        Viscosity is the resistance offered by a fluid to its flow. A high vis-
cosity fluid has molecules which move over each other only with great
difficulty in effecting a change of shape. The viscosity of fluids must be
compared at the same temperature. (For example, molasses is viscous
and flows more easily in summer than in winter.) A measurement of the
viscosity of oil, sometimes given in Saybolt seconds, or just Saybolt, is
the time required for a sample of oil to flow through a standard orifice
at a standard temperature of 122°F; the instrument used for determining
the viscosity is known generally as a Saybolt viscosimeter. A knowledge
of the viscosity is very important as oil must be pumped through pipe
lines and small orifices of burners.

Flash Point
        Flash point is the temperature at which oil begins to give off flam-
mable gas in sufficient quantities to ignite. The minimum temperature,

for reasons of safety, is 150°F and is important in connection with the storage of oil. The flammable vapor collecting in a closed space such as the top of a storage tank may be the cause of an explosion if an open flame or spark is present when the tank is opened and air allowed to mix with the vapor.

Heating Value

Heating value of oil is the number of Btu's liberated by the perfect combustion of one pound of oil. This may be determined experimentally by burning a weighed sample of oil in a calorimeter. The heating value of many fuel oils varies from 18,000 to 19,500 Btu per pound. Some standard properties of Bunker C fuel oil as found in some specifications are:

Specific Gravity .... 5 to 14 degrees API

Viscosity .... 300 seconds Saybolt at 122°F

Btu .... Approx. 18,500 Btu per pound

Water Content .... Not more than 2.0 percent

Sediment Content .... Not more than 0.25 percent

Combustion Process

The basic chemical combination process of fuel oil is similar to that of pulverized coal; indeed, fuel oil may well be considered as almost another form of pulverized coal having approximately 100 percent volatile matter content.

At high temperatures, the oil is vaporized and forms rich hydrocarbon gases which ignite and burn by combining with oxygen from the air supply, just as is the case with the volatile matter of coal. Thus the burning of oil follows the laws of gases in that the molecules of the hydrocarbons in gaseous form meet the molecules of oxygen. Combustion will be complete provided there is an intimate mixing of the gaseous volatiles and oxygen.

For the efficient combustion of fuel oil, conditions similar to those required for the combustion of pulverized coal are necessary. These are:

1. Complete atomization.
2. Temperature sufficient to maintain rapid and continuous ignition.
3. Sufficient air supply - excess air.
4. Adequate mixing of air and oil.

As in pulverized coal, the ignition temperature is that of the volatile hydrocarbons. In fuel oil, however, there is always present a certain amount of high volatile hydrocarbons whose ignition temperature is quite low, which are assets to the maintenance of ignition.

To insure complete combustion, excess air is necessary; if there is insufficient excess air the heavier hydrocarbons will "crack" or break down, producing soot and oily tar vapors which are difficult to burn. Even though excess air may be present, cracking or delayed ignition may result if atomization is not complete and satisfactory turbulence has not been obtained. This is necessary if excess air is to be reduced to a minimum and stable ignition maintained.

Heating of Fuel Oil

Fuel oil is heated for the sole purpose of reducing its viscosity. This is necessary to facilitate pumping and to secure a fineness of atomization. A low viscosity will gives

1. Finer atomization.
2. More rapid and efficient combustion.
3. Greater rate of combustion.

Heating above 220°F is, however, never attempted as many commercial fuels will carbonize and precipitate sludge at that value.

Oil Ash

Oil ash may contain a large number of oxides in varying but very small quantities that include: silicon (SiO_2); iron (FeO_2); aluminum (Al_2O_2); calcium (CaO); magnesium (MgO); sulphur (SO_2): nickel (NiO); vanadium (V_2O_2); sodium (NaO); and lesser quantities of compounds of phosphorus, copper, tin, lead and other elements. Vanadium, classified as a rare element, and used in the hardening of steels, is economically recoverable.

NATURAL GAS

Natural gas almost always accompanies petroleum and is released when the pressure on the oil is released. The gas consists mainly of light hydrocarbons such as methane (CH_4), ethane (C_2H_6), propane (C_3H_8)j butane (C_4H_{10}), and others, all of which are volatile. Propane and butane are often separated from the lighter gases and are liquefied under pressure for transportation as liquids; when the pressure is released, the liquid boils producing gaseous fuel.

Much of the natural gas produced is suitable for use without further preparation. Some natural gas, however, may contain enough sand or gaseous sulphur to be troublesome, and these are removed by "scrubbing" processes which filter out the solids and change the sulphur into recoverable sulfides and sulfuric acid. Natural gas may be diluted with air or enriched by oil sprays to keep its Btu content per cubic foot essentially constant.

Of all fuels, natural gas is considered the most desirable for power generation. Often there is no need for its storage. It is substantially free of ash; combustion is smokeless, and, because it is a gas, it mixes easily and intimately with air to give complete combustion with low excess combustion air. Although the total hydrogen content of natural gas is relatively high, the amount of free hydrogen is low. Because of the high hydrogen content as compared to coal and oil, the water vapor formed in combustion is correspondingly greater, which may be reflected in lower efficiency of combustion. As with all gaseous fuels, its ease of mixing with air increases the hazards of explosion, requiring greater care in its handling.

OTHER FUELS

Fuels, other than coal, oil and natural gas, may include such substances as:

- Vegetation, both land and aquatic; e.g. bagasse from sugar cane, corn husks, etc.

- Tars as residue from coal and coke operations; e.g. from manufactured gas, steel making, etc.

- Garbage from industrial, commercial and residential waste, etc.

- Gases from garbage piles and manure heaps, etc.

The conversion of energy from geothermal sources, wind, direct solar exposure, tides, temperature differences between surface and deep sections of bodies of water and other esoteric sources, are all in the experimental stage whose economic feasibility is under consideration. See Chapter 9.

Of all the sources mentioned, only hydro, coal, oil and natural gas are subject to procedures that may be considered adaptable to standardized operations. The handling and conversion of the others are subject to local considerations as to their practicability and conversion processes, and are beyond the scope of this work. Nuclear fuel will be discussed separately.

COMPARISON OF FUELS

The decision as to what fuel should be burned depends on the individual plant, the equipment and operating personnel, local conditions, and principally on the price of the fuel. The price, in turn, may be affected by its availability, transportation and handling costs.

In very small plants, hand firing with its consequent inefficient operation, cannot compete with oil or gas firing, but higher prices for oil and gas, and lower labor costs, may make such firing competitive. In large central generating plants, the method and fuel used, whether stoker, pulverized coal, oil or gas, vary so little that price of the fuel is usually the determining factor.

Some of the advantages and disadvantages of oil and gas over coal are listed below; those for oil and gas are essentially the same, except those for gas are usually greater than for oil.

Advantages
1. Reduction in fuel handling costs.
2. Labor savings; no ash or dust removal, etc.
3. Reduction, or elimination, of storage space.
4. High efficiencies and capacities (especially for small plants).
5. No loss in heat value when properly stored.
6. Cleanliness and freedom from dust.

Disadvantages
1. Usually cost more on equal Btu basis; gas, oil, coal in descending order.
2. Danger from explosion if not properly handled, gas more than oil.
3. Furnace maintenance high; may require closer supervision.

DISTRIBUTED GENERATION

As the name implies, generating units of relatively small dimensions and capacity (from 5 kW to 5 MW) are installed at or near the load to be served and usually connected to the associated distribution or transmission system. These units can serve:

1. to supply base load

2. for peak shaving

3. to supply additional load instead of revamping of existing supply facilities

4. as independent producers of electricity in areas remote from system sources, or where it may be more economic to do so

These units share some similarity with cogeneration but are not necessarily installed to take advantage of available steam associated with heating, cooling or other process loads. These units may be powered by reciprocating steam or internal combustion engines, by small gas or steam turbines (micro turbines), or may be fuel cells. Their protective systems are essentially the same as those for cogeneration. Improper operation of the protective devices may result in the energizing of lines and equipment presumed de-energized and on which workers may be working.

Like the transmission lines, the questions arise as to who will build, own, operate and maintain them, a most important consideration as inadequate systems and maintenance may result from insufficiently informed personnel, and this is especially true regarding the protective systems and devices.

The possibility of "islanding," a condition to which a distributed generation energizes a portion of a distribution system at a time when the remainder of the systems de-energized unintentionally occurs, can result in safety hazards and damage to the consumer's system and that of the utility to which it is connected.

Typical requirements for connecting new Distributed Generation units to the utility system are included below.

REQUIREMENTS FOR INTERCONNECTION OF
NEW DISTRIBUTED GENERATION UNIT WITH
CAPACITY OF 300 kVA OR LESS TO BE OPERATED IN
PARALLEL WITH RADIAL DISTRIBUTION LINES*

A. Design Requirements

1. Common

The generator-owner shall provide appropriate protection and control equipment, including an interrupting device, that will disconnect[1] the generation in the event that the portion of the _____ system that serves the generator is de-energized for any reason or for a fault in the generator-owner's system. The generator-owner's protection and control equipment shall be capable of disconnecting the generation upon detection of an islanding[2] condition and upon detection of a _____ system fault.

The generator-owner's protection and control scheme shall be designed to allow the generation, at steady state, to operate only within the limits specified in this document for frequency and voltage. Upon request from _____, the generator-owner shall provide documentation detailing compliance with the requirements set forth in this document.

The specific design of the protection, control and grounding schemes will depend on the size and characteristics of the generator-owner's generation, as well the generator owner's load level, in addition to the characteristics of the particular portion of _____ system where the generator-owner is interconnecting.

The generator-owner shall have, as a minimum, an interrupting device(s) sized to meet all applicable local, state and federal codes and operated by over and under voltage protection on each phase. The interrupting device(s) shall also be operated by over and under frequency protection on at least one phase. All phases of a generator or inverter interface shall disconnect for a voltage or frequency trip on any phase. It is recommended that voltage protection be wired phase to ground.

---

*Courtesy of LILCO
[1]See Glossary for definition
[2]Ibid

— The interrupting device shall automatically initiate a disconnect sequence from the _____ system within six (6) cycles if the voltage falls below 60 V rms phase to ground (nominal 120 V rms base) on any phase.

— The interrupting device shall automatically initiate a disconnect sequence from the _____ system within two (2) seconds if the voltage rises above 132 V rms phase to ground or falls below 106 V rms phase to ground (nominal 120 V rms base) on any phase.

— The interrupting device shall automatically initiate a disconnect sequence from the _____ system within two (2) cycles if the voltage rises above 165 V rms phase to ground (nominal 120 V rms base) on any phase.

— The interrupting device shall automatically initiate a disconnect sequence from the _____ system within six (6) cycles if the frequency rises above 60.5 Hz or falls below 59.3 Hz.

The need for additional protection equipment shall be determined by _____ on a case-by-case basis. The _____ shall specify and provide settings for those relays that _____ designates as being required to satisfy protection practices. Any protective equipment or setting specified by _____ shall not be changed or modified at any time by the generator-owner without written consent from _____.

To avoid out-of-phase reclosing, the design of the generator-owner's protection and control scheme shall take into account _____'s practice of automatically reclosing the feeder without synchronism check as quickly as 12 cycles after being tripped.

The generator-owner shall be responsible for ongoing compliance with all applicable local, state and federal codes and standardized interconnection requirements as they pertain to the interconnection of the generating equipment.

Protection shall not share electrical equipment associated _____ revenue metering.

A failure of the generator-owner's interconnection protection equipment, including loss of control power, shall open the interrupting device, thus disconnecting the generation from the _____ system. A generator-owner's protection equipment shall utilize a nonvolatile memory design such that a loss of internal or external control power, including batteries, will not cause a loss of interconnection protection functions or loss of protection set points.

All interface protection and control equipment shall operate as specified independent of the calendar date.

2. Synchronous Generators

Synchronous generation shall require synchronizing facilities. These shall include automatic synchronizing equipment or manual synchronizing with relay supervision, voltage regulator and power factor control.

3. Induction Generators

Induction generation may be connected and brought up to synchronous speed (as an induction motor) if it can be demonstrated that the initial voltage drop measured at the point of common coupling (PCC)[3] is acceptable based on current inrush limits. The same requirements also apply to induction generation connected at or near synchronous speed because a voltage dip is present due to an inrush magnetizing current. The generator-owner shall submit the expected number of starts per specific time period and maximum starting kVA draw data to _____ to verify that the voltage dip due to starting is within the visible flicker limits as defined by IEEE 519-1992, Recommended Practices and Requirements for Harmonic Control in Electric Power Systems (IEEE 519).

Starting or rapid load fluctuations on induction generators can adversely impact _____'s system voltage. Corrective step-switched capacitors or other techniques may be necessary. These measures can, in turn, cause ferroresonance. If these measures (additional capacitors) are installed on the customer's side of the PCC, _____ will review these measures and may require the customer to install additional equipment.

_____

3. See Glossary for definition

4. DC Inverters

Direct current generation can only be installed in parallel with _____ ___'s system using a synchronous inverter. The design shall be such as to disconnect this synchronous inverter upon a _____ system interruption.

Line-commutated inverters do not require synchronizing equipment if the voltage drop is determined to be acceptable, as defined in Section IV(E), Power Quality, of this document. Self-commutated inverters of the utility-interactive type shall synchronize to _____. Stand-alone, self-commutated inverters shall not be used for parallel operation with _____.

A line inverter can be used to isolate the customer from the _____ system provided it can be demonstrated that the inverter isolates the customer from the _____ system safely and reliably.

Voltage and frequency trip set points for inverters shall be accessible to service personnel only.

5. Metering

The need for additional revenue metering or modifications to existing metering will be reviewed on a case-by-case basis and shall be consistent with metering requirements adopted by _____.

B. Operating Requirements

The generator-owner shall provide a 24-hour telephone contact(s). This contact will be used by _____ to arrange access for repairs, inspection or emergencies. _____ will make such arrangements (except for emergencies) during normal business hours.

The generator-owner shall not supply power to _____ during any outages of the system that serves the PCC. The generator-owner's generation may be operated during such outages only with an open tie to _____. Islanding will not be permitted. The generator-owner shall not energize a de-energized _____ circuit for any reason.

Generation that does not operate in parallel with the _____ system is not subject to these requirements.

The disconnect switch[4] specified in Section IV(D) of this document may be opened by _____ at any time for any of the following reasons:

a. To eliminate conditions that constitute a potential hazard to _____ __ personnel or the general public;
b. Pre-emergency or emergency conditions on the _____ system;
c. A hazardous condition is revealed by a _____ inspection;
d. Protective device tampering.

The disconnect switch may be opened by _____ for the following reasons, after notice to the responsible party has been delivered and a reasonable time to correct (consistent with the conditions) has elapsed:

a. A power producer has failed to make available records of verification tests and maintenance of its protective devices;
b. A power producer's system interferes with 40M equipment or equipment belonging to other _____ customers;
c. A power producer's system is found to affect quality of service of adjoining customers.

_____ will provide a name and telephone number so that the customer can obtain information about the _____ lock-out. The customer shall be allowed to disconnect from _____ without prior notice in order to self-generate.

Following a generation facility disconnect as a result of a voltage or frequency excursion, the generation facility shall remain disconnected until _____'s service voltage and frequency has recovered to _____ ____'s acceptable voltage and frequency limits for a minimum of five (5) minutes.

_____ may require direct transfer trip (DTT)[5] whenever: 1) the minimum load to generation ratio on a circuit is such that a ferroresonance condition could occur; 2) it is determined that the customer's protective

---

[4]See Glossary for definition.
[5]Ibid

relaying may not operate for certain conditions or faults and/or 3) the installation could increase the length of outages on a distribution circuit or jeopardize the reliability of the circuit. _____ will be required to demonstrate the need for DTT.

C. Dedicated Transformer[6]

_____ reserves the right to require a power producing facility to connect to the _____ system through a dedicated transformer. The transformer shall either be provided by _____ at the generator-owner's expense, purchased from _____, or provided by the generator owner in conformance with _____'s specifications. The transformer may be necessary to ensure conformance with _____ safe work practices, to enhance service restoration operations or to prevent detrimental effects to other _____ customers. The dedicated transformer that is part of the normal electrical service connection of a generator-owner's facility may meet this requirement if there are no other customers supplied from it. A dedicated transformer is not required if the installation is designed and coordinated with _____ to protect the _____ system and its customers adequately from potential detrimental net effects caused by the operation of the generator.

If _____ determines a need for a dedicated transformer, it shall notify the generator owner in writing of the requirements. The notice shall include a description of the specific aspects of the _____ system that necessitate the addition, the conditions under which the dedicated transformer is expected to enhance safety or prevent detrimental effects, and the expected response of a normal, shared transformer installation to such conditions.

D. Disconnect Switch

Generating equipment shall be capable of being isolated from the _____ system by means of an external, manual, visible, gang-operated, load break disconnecting switch. The disconnect switch shall be installed, owned and maintained by the owner of the power producing facility and located between the power producing equipment and its interconnection point with the _____ system.

---

[6]See Glossary for definition.

The disconnect switch must be rated for the voltage and current requirements of the installation.

The basic insulation level (BIL) of the disconnect switch shall be such that it will coordinate with that of _____'s equipment. Disconnect devices shall meet applicable UL, ANSI and IEEE standards, and shall be installed to meet all applicable local, state and federal codes. (City Building Code may require additional certification.)

The disconnect switch shall be clearly marked, "Generator Disconnect Switch," with permanent 3/8-inch letters or larger.

The disconnect switch shall be located within 10 feet of _____'s external electric service meter, or the location and nature of the distributed power disconnection switches shall be indicated in the immediate proximity of the electric service entrance.

The disconnect switch shall be readily accessible for operation and locking by _____ personnel in accordance with Section IV(B) of this document.

The disconnect switch must be lockable in the open position with a standard _____ padlock with a 3/8-inch shank.

E. Power Quality

The maximum harmonic limits for electrical equipment shall be in accordance with IEEE 519. The objective of IEEE 519 is to limit the maximum individual frequency voltage harmonic to 3% of the fundamental frequency and the voltage Total Harmonic Distortion (THD) to 5% on the _____ side of the PCC. In addition, any voltage flicker resulting from the connection of the customer's energy producing equipment to the _____ system must not exceed the limits defined by the maximum permissible voltage fluctuations border line of visibility curve, Figure 10.3 identified in IEEE 519. This requirement is necessary to minimize the adverse voltage effect upon other customers on the _____ system.

F. Power Factor

If the power factor, as measured at the PCC, is less than 0.9 (leading or lagging), the method of power factor correction necessitated by the installation of the generator will be negotiated with _____ as a commercial item.

Induction power generators may be provided VAR capacity from the ____ _____ system at the generator-owner's expense. The installation of VAR correction equipment by the generator-owner on the generator-owner's side of the PCC must be reviewed and approved by _____ prior to installation.

G. Islanding

Generation interconnection systems must be designed and operated so that islanding is not sustained on radial distribution circuits. The requirements listed in this document are designed and intended to prevent islanding.

H. Test Requirements

This section is divided into type testing and verification testing. Type testing is performed or witnessed once by an independent testing laboratory for a specific protection package. Once a package meets the type test criteria described in this section, the design is accepted by _____. If any changes are made to the hardware, software, firmware, or verification test procedures, the manufacturer must notify the independent testing laboratory to determine what, if any, parts of the type testing must be repeated. Failure of the manufacturer to notify the independent test laboratory of changes may result in withdrawal of approval and disconnection of units installed since the change was made. Verification testing is site-specific, periodic testing to assure continued acceptable performance.

Type testing results shall be reported to the State Department of Public Service. Department Staff shall review the test report to verify all the appropriate tests have been performed. The Department of Public Service will maintain a list of equipment that has been type tested and approved for interconnection in the State. The list will contain discrete protective relays as well as inverters with integrated protection and control. The list will indicate specific model numbers and firmware versions approved. The equipment in the field must have a nameplate that clearly shows the model number and firmware version (if applicable).

These test procedures apply only to devices and packages associated with protection of the interface between the generating system and the

_____ system. Interface protection is usually limited to voltage relays, frequency relays, synchronizing relays, reverse current or power relays, and anti-islanding schemes. Testing of relays or devices associated specifically with protection or control of generating equipment is recommended, but not required unless they impact the interface protection.

At the time of production, all interconnecting equipment including inverters and discrete relays must meet or exceed the requirements of ANSI IEEE C62.41-1991 -Recommended Practices on Surge Voltages in Low Voltage AC Power Circuits or C37.90.1 1989, IEEE Standard Surge Withstand Capability (SEC) Tests for Protective Relays and Relay Systems. If C62.41-1991 is used, the surge types and parameters shall be applied, as applicable, to the equipment's intended insulation location. If the device is not tested to level C voltage, i.e., for an intended location on the _____ side of the meter, the test report shall record the voltage level to which the device was tested and the Public Service Commission listing shall specify the location limitations of the device.

If after the application of the surge test, the unit is still functioning and has the capability to export power to the _____ system, it shall be subjected to and comply with the manufacturer's verification test and the appropriate dielectric test as specified in UL 1741.

All single-phase and three phase test voltages shall be applied phase to ground.[7]

Isolation transformers specified as required or listed as optional must be connected. Each optional isolation transformer connection constitutes a separate type test. Generic isolation transformers may be substituted after type testing.

Three-phase isolation transformers connected wye-grounded/delta on the generator side are not permitted.

---

[7]Test voltages are specified phase to ground for a 120 volt nominal system. Other system voltages require adjusting the test voltages by the appropriate percentage. Over and undervoltage protection should be wired phase to ground. Phase to phase voltage sensing results in less sensitive undervoltage detection and more sensitive overvoltage detection.

1. Type Testing

All interface equipment must include a verification test procedure as part of the documentation. Except for the case of small single-phase inverters discussed below, the verification test must determine if protection settings meet these requirements. The independent testing laboratory shall conduct the verification test prescribed by the manufacturer to determine if the verification test procedure adequately demonstrates compliance with these requirements.

Prior to testing, all batteries shall be disconnected or removed for a minimum of ten (10) minutes. This test is to verify the system has a non-volatile memory and that protection settings are not lost. A test shall also be performed to determine that failure of any battery not used to supply trip power will result in an automatic shutdown.

a. Single-Phase Inverters

All single-phase inverters shall be non-islanding inverters as defined by IEEE P929. Inverters 1OkW and below shall at the time of production meet or exceed the requirements of IEEE 929 and UL 1741. Specifically, the inverter shall automatically disconnect for an islanding condition with load quality factor of 2.5 within two (2) seconds. In addition, all single-phase inverters and single-phase voltage and frequency relay packages shall initiate a trip from a waveform generator for the waveforms listed below to verify they meet the requirements set forth in the design section of this document.

Waveform 1- A 120 V rms 60 Hz sinusoidal that drops in voltage to 59 V rms for six (6) cycles beginning and ending at a zero crossing and resuming to 120 V rms for five minutes.

Waveform 2- A 120 V rms 60 Hz sinusoidal that drops in voltage to 105 V rms for 120 cycles beginning and ending at a zero crossing and resuming to 120 V rms for five minutes.

Waveform 3- A 120 V rms 60 Hz sinusoidal that rises in voltage to 133 V rms for 120 cycles beginning and ending at a zero crossing and resuming to 120 V rms for five minutes.

Waveform 4- A 120 V rms 60 Hz sinusoidal that rises in voltage to 166 volts for two (2) cycles beginning and ending at a zero crossing and re-suming to 120 V rms for five minutes.

Waveform 5- A 120 V rms 60 Hz sinusoidal that drops in frequency to 59.2 Hz for six (6) cycles beginning and ending at a zero crossing and resuming to 60 Hz for five minutes.

Waveform 6- A 120 V rms 60 Hz sinusoidal that rises in frequency to 60.6 Hz for six (6) cycles beginning and ending at a zero crossing and resuming to 60 Hz for five minutes.

Each waveform test shall be repeated ten (10) times. Failure to cease to ex-port power for any one run constitutes failure of the test. These tests shall also verify the inverter or power producing facility shall not automatically reconnect to the waveform generator until after five (5) minutes of continu-ous normal voltage and frequency. The manufacturer may supply a special production sample with the five minute reset timer disabled to eliminate waiting time during type testing. At least one test must be performed on a sample with a five minute reset timer to verify the function and accuracy of the timer.

b. Three-Phase Inverters

Three-phase inverters and discrete three-phase voltage relays shall be type tested with three phase waveforms. The inverter shall disconnect or the protection equipment shall initiate a trip from the waveform generator for each of the waveforms described below:

Waveform 1 - A three-phase sinusoidal operating at 60 Hz and 120 V rms interrupted by phase A voltage depressed to 59 V rms for six (6) cycles beginning and ending at a zero crossing while B and C phases continue at 120 V rms. Repeat the same test with B phase depressed, with C phase depressed, with A and B phases depressed, with B and C phases de-pressed, and finally with all phases depressed to 59 V for six cycles.

Waveform 2- A three-phase sinusoidal operating at 60 Hz and 120 V rms interrupted by phase A voltage depressed to 59 V rms for six (6) cycles beginning and ending at a zero crossing while B and C phases are in-creased to 150 V rms beginning and ending at the same point of discon-tinuity. Repeat the same test with B phase depressed and A and C phases increased and with C phase depressed and A and B phases increased.

Waveform 3- A three-phase sinusoidal operating at 60 Hz and 120 V rms interrupted by phase A voltage depressed to 105 V rms for two seconds (120 cycles) beginning and ending at a zero crossing while B and C phases continue at 120 V rms. Repeat the same test with B and C phases depressed to the same level and for the same duration.

Waveform 4- A three-phase sinusoidal operating at 60 Hz and 120 V rms interrupted by phase A voltage increased to 133 V rms for two seconds (120 cycles) beginning and ending at a zero crossing while B and C phases continue at 120 V rms. Repeat the same test with B and C phases increased to the same level and for the same duration.

Waveform 5- A three-phase sinusoidal operating at 60 Hz and 120 V rms interrupted by phase A voltage increased to 166 V rms for two seconds (120 cycles) beginning and ending at a zero crossing while B and C phases continue at 120 V rms. Repeat the same test with B and C phases increased to the same level and for the same duration.

Waveform 6- A three-phase sinusoidal operating at 60 Hz and 120 V rms interrupted by phase A voltage increased to 166 V rms for two cycles beginning and ending at a zero crossing while B and C phases are decreased to 100 V rms beginning and ending at the same point of discontinuity. Repeat the same test with B phases increased and A and C phases decreased and for C phase increased and A and B phases decreased to the same levels and for the same duration.

Waveform 7- A three phase sinusoidal operating at 60 Hz and 120 V rms interrupted with six (6) cycles of 59.2 Hz beginning and ending at the zero crossing on A phase.

Waveform 8- A three-phase sinusoidal operating at 60 Hz and 120 V rms interrupted with six (6) cycles of 59.2 Hz beginning and ending at the zero crossing on B phase and with A and C phase voltages depressed to 70 V rms beginning and ending at the same point of discontinuity.

Waveform 9- A three-phase sinusoidal operating at 60 Hz and 120 V rms interrupted with six (6) cycles of 60.6 Hz beginning and ending at the zero crossing on A phase.

Waveform 10- A three-phase sinusoidal operating at 60 Hz and 120 V rms interrupted with six (6) cycles of 60.6 Hz beginning and ending at the zero crossing on C phase and with A and B phase voltage depressed to 70 V rms beginning and ending at the same point of discontinuity.

Each three-phase waveform test shall be repeated ten (10) times. Failure to trip for anyone run. constitutes failure of the test. These tests shall also verify the inverter or power producing facility shall not automatically reconnect to the waveform generator until after five (5) minutes of continuous normal voltage and frequency. The manufacturer may supply a special production sample with the five minute reset timer disabled to eliminate waiting time during type testing. At least one test must be performed on a sample with a five minute reset timer to verify the function and accuracy of the timer.

Alternatively, three-phase inverters with integrated protection and control may be tested with a generator to simulate abnormal _____ frequency and voltages. Abnormal _____ voltage may also be simulated with an autotransformer/variac. The tests shall include:

Test 1: With the generator and inverter output stabilized at 60 Hz and 120 V rms and the inverter output between 0.5 and 1.0 per unit power, ramp the generator voltage up to 133 V rms at a rate no greater than 5 volts per second. Measure and record the frequency and voltage. The frequency must remain within 0.2 Hz of 60 Hz and the voltage may not exceed 137 V rms. The inverter must cease to export power within two seconds (120 cycles) of the first half-cycle reaching 188 V peak to neutral. Repeat the test with the inverter output below 0. 1 per unit power.

Test 2: Insert a tapped transformer and a breaker between A phase of the generator and A phase of the inverter arranged such that when the breaker is opened or closed, A phase of the inverter receives half the voltage of the generator. With the generator and inverter output stabilized at 60 Hz and 119 V rms and the inverter output between 0.5 and 1.0 per unit power, operate the breaker so A phase of the inverter only receives 58 V rms. Measure and record the frequency and voltage. The frequency must remain within 0.2 Hz of 60 Hz and the voltage may not drop below 55 V rms on A phase of the inverter or below 110 V rms on B or C phases of the inverter. The inverter must cease to export power within six cycles of when the first half cycle of voltage on A phase of the inverter drops below 83 V peak to neutral. Repeat the test applying half voltage to B and C phases. And repeat the test for all phases with the inverter output below 0.1 per unit power.

Test 3: With the generator and inverter output stabilized at 60 Hz and 120 V rms and the inverter output between 0.5 and 1.0 per unit power, ramp the generator voltage down to 103 V rms at a rate no greater than 5 volts per second. Measure and record the frequency and voltage. The frequency must remain within 0.2 Hz of 60 Hz and the voltage must not drop below 99 V rms. The inverter must cease to export power within two seconds (120 cycles) of the first half-cycle reaching 145 V peak to neutral. Repeat the test with the inverter output below 0. 1 per unit power.

Test 4: Insert a tapped transformer and a breaker between A phase of the generator and A phase of the inverter arranged such that when the breaker is opened or closed, A phase of the inverter receives four-fifths the voltage of the generator. With the generator and inverter output stabilized at 60 Hz and 128 V rms and the inverter output between 0.5 and 1.0 per unit power, operate the breaker so that A phase of the inverter only receives 103 V rms. Measure and record the frequency and voltage. The frequency must remain within 0.2 Hz of 60 Hz and the voltage may not drop below 99 V rms on A phase of the inverter, or below 110 V rms on B or C phases of the inverter. The inverter must cease to export power within two seconds (120 cycles) of when the first half cycle of voltage on A phase of the inverter drops below 145 V peak to neutral. Repeat the test applying low voltage to B and C phases. And repeat the test for all phases with the inverter output below 0.1 per unit power.

Test 5: With the generator and inverter output stabilized at 60 Hz and 120 V rms and the inverter output between 0.5 and 1.0 per unit power, ramp the generator frequency up to 60.6 Hz at a rate no greater than 0.5 Hz per second. Measure and record the frequency and voltage. The voltage must remain between 115 V rms and 125 V rms and the frequency must not exceed 60.8 Hz. The inverter must cease to export power within six cycles of the frequency exceeding 60.6 Hz (8.25 ms between zero crossings). Repeat the test with the inverter output below 0.1 per unit power.

Test 6: With the generator and inverter output stabilized at 60 Hz and 120 V rms and the inverter output between 0.5 and 1.0 per unit power, ramp the generator frequency down to 59.2 Hz at a rate no greater than 0.5 Hz per second. Measure and record the frequency and voltage. The voltage must remain between 115 V rms and 125 V rms and the frequency must not fall below 59.0 Hz. The inverter must cease to export power within six cycles of the frequency falling below 59.2 Hz (8.22 ms between zero crossings). Repeat the test with the inverter output below 0.1 per unit power.

Tests 1 through 6 above shall be repeated five (5) times. Failure to cease to export power for any one run where the frequency and voltage are recorded and fall outside of the accepted limits shall constitute failure of the test. Following at least one run of each test group, the generator is to remain running to verify that the inverter does not automatically reconnect until after five (5) minutes of continuous normal voltage and frequency.

It is not necessary to perform the 165 V rms test, the 132 V rms unbalanced voltage test, or the anti-islanding test on three phase inverters.

2. Verification Testing

Upon initial parallel operation of a generating system, or any time interface hardware or software is changed, a verification test must be performed. A licensed professional engineer or otherwise qualified individual must perform verification testing in accordance with the manufacturer's published test procedure. Qualified individuals include professional engineers, factory trained and certified technicians, and licensed electricians with experience in testing protective equipment. _____ reserves the right to witness verification testing or require written certification that the testing was performed.

Verification testing shall be performed every four years. All verification tests prescribed by the manufacturer shall be performed. If wires must be removed to perform certain tests, each wire and each terminal must be clearly and permanently marked. The generator-owner shall maintain verification test reports for inspection by _____.

Single-phase inverters rated 15 kVA and below may be verified once per year as follows: once per year, the owner or his agent shall operate the load break disconnect switch and verify the power producing facility automatically shuts down and does not restart for five minutes after the switch is closed. The owner shall maintain a log of these operations for inspection by _____.

Any system that depends upon a battery for trip power shall be checked and logged once per. month for proper voltage. Once every four (4) years the battery must be either replaced or a discharge test performed.

Glossary of Terms:

Automatic Disconnect Device An electronic or mechanical switch used to isolate a circuit or piece of equipment from a source of power without the need for human intervention.

Coordinated Interconnection Review - Any studies performed by utilities to ensure that the safety and reliability of the electric grid with respect to the interconnection of distributed generation as discussed in this document.

Dedicated Service Transformer or Dedicated Transformer A transformer with a secondary winding that serves only one customer.

Direct Transfer Trip (DTT) - remote operation of a circuit breaker by means of a communication channel.

Disconnect (verb) - to isolate a circuit or equipment from a source of power. If isolation is accomplished with a solid state device, "Disconnect" shall mean to cease the transfer of power.

Disconnect Switch A mechanical device used for isolating a circuit or equipment from a source of power.

Energy Conversion Device A machine or solid state circuit for changing direct current to alternating current or a machine that changes shaft horse-power to electrical power.

Islanding A condition in which a portion of the _____ system that contains both load and distributed generation is isolated from the remainder of the _____ system. [Adopted from IEEE 929, draft 9].

Point of Common Coupling (PCC) The point at which _____ and the customer interface occurs. Typically, this is the customer side of the _____ revenue meter. [Adopted from IEEE 929, draft 9].

Radial Feeder A distribution line that branches out from a substation and is normally not connected to another substation or another circuit sharing the common supply.

Type Test - A test performed or witnessed once by a qualified independent testing laboratory for a specific protection package or device to determine whether the requirements of this document are met. The Type Test will typically be sponsored by equipment manufacturers.

Verification Test - A test performed upon initial installation and repeated periodically to determine that there is continued acceptable performance.

NUCLEAR FUEL - NUCLEAR REDUCTION

A nuclear fuel power plant differs from a fossil fuel power plant in that a nuclear reactor and a specialized boiler are substituted for the conventional furnace and boiler. The major difference is found in the handling of nuclear fuel and boiler room equipment. Steam still plays the dominant role in the production of electricity. See Figure 2-2.

Nuclear Energy

Nuclear energy (sometimes less precisely referred to as atomic energy), may be defined as the energy which is created when mass is destroyed. The equivalence of matter and energy are expressed in Einstein's well known equation:

$$E = mc^2$$

Since the velocity of light (c) is 186,000 miles per second, a very small change in mass (m) will produce an enormous amount of energy. For one pound of mass, the energy released will be found to be:

$$E/lb = 38,690 \; billion \; Btu$$

or (assuming 13,000 Btu per pound of coal) some three billion times as much energy in uranium (pound for pound)! However, a way has not yet been found to convert all of a given bulk of matter to energy. Even in the normal reaction (or fission) of uranium, only about one-thousandth of the mass is consumed. So when one pound of uranium is fissioned, some three-million times more energy is obtained than by burning one pound of coal. A fantastic potential!

(To put nuclear energy into proper perspective, while Einstein determined the relationship between energy and mass, he also was convinced that energy derived from mass was impossible. Others determined that it was possible, and it remained for Fermi to demonstrate not only its possibility, but its practicability as well.)

Nuclear Fuel

In describing the combustion of fossil fuels, the atoms of one substance (e.g. carbon) combine with the atoms of another (e.g. oxygen) to form a molecule of a third (e.g. carbon dioxide), liberating heat in the process.

In nuclear fuel, energy is released by reactions that take place within the atom itself. The atom, as small as it is, is made up of still smaller particles and the magnitudes of the spaces between them bear the same relative relationship as the elements of the solar system; indeed, the atom can be portrayed as a miniature solar system.

At the center of this system is a *nucleus*. Around this small, but relatively heavy center, particles having a negative electric charge, called electrons, spin at very high speeds. The nucleus is made up of two kinds of particles, protons and neutrons. Protons are positively charged particles and usually are equal in number to the electrons. Neutrons resemble the protons but carry no electrical charge. Practically all of the atom's mass is in the nucleus; one electron has only about one two-thousandth (0.002) the mass of the proton. To give some idea of the minuteness of relative dimensions involved, if the nucleus was as large as a baseball, electrons would be specks a half-mile away. The diameter of an atom, which is also the electron orbit, is in the nature of two one-hundred-millionths (0.000,000,002) of an inch, and the diameter of the nucleus about one ten-thousandth (0.0001) of the diameter of the atom, or two trillionths (0.000,000,000,002) of an inch. See Figure 2-7.

Nuclear Reaction

In the combustion of fossil fuel, a chemical reaction takes place in which the electrons of the carbon and oxygen atoms undergo rearrangements that release heat, light, and combustion products.

In the nuclear reaction, the nucleus of the atom is involved rather than the electrons. The forces holding together the protons and neutrons are considerably greater than the forces that hold the electrons in their orbits. Hence, the energy released when a neutron is split away from the nucleus is enormously greater than when oxygen and carbon atoms are combined.

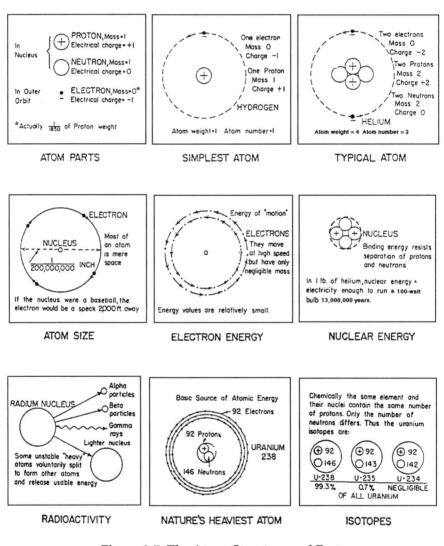

Figure 2-7. The Atom: Structure and Parts
(Courtesy National Industrial Conference Board)

While theoretically neutrons can be split away from the nucleus of any substance, practically the choice is limited only to a few of the elements that have a relatively large number of neutrons in their nuclei, that are readily available, and comparatively easy to be separated from their compounds, purified, and worked into some convenient form for use in a reactor. Uranium is practically alone in meeting these requirements.

Uranium Isotopes

Natural uranium contains three types of uranium, distinguished by their "isotopes." Isotopes are the same chemical element whose nuclei have the same number of protons but a different number of neutrons:

Isotopes	Percent of Uranium	Protons	Neutrons	Total (Atomic Mass)
Uranium-234	0.006	92	142	234
Uranium-235	0.712	92	143	235
Uranium-238	99.282	92	146	238

Of these three isotopes, only uranium-235 is readily and naturally fissionable, that is, its nucleus readily "splits" or ejects its neutrons, releasing energy mainly in the form of heat.

Nuclear Fission

The fission process and chain reaction are fundamental to the operation of a nuclear reactor.

Since the neutron has no electrical charge, it can penetrate into the nucleus of an atom without being affected by the negatively charged electrons or positively charged protons. If a neutron enters a mass of uranium and it strikes the nucleus of an atom of its isotope, uranium-235, it is probable that the nucleus will split (or fission), releasing the large amount of energy that holds the nucleus together. In addition to generating heat in the mass of uranium, and emitting radiation, the fissioned nucleus will also eject one or more neutrons at high speed which, in turn, will hit and fission other nuclei in the uranium mass. The neutron "bullets" act somewhat as the oxygen atoms do when combining with the atoms of carbon, producing heat, light and waste products. See Figure 2-8.

Since the isotope uranium-235 constitutes only 0.7 percent or one part in 140 of natural uranium, and some neutrons are lost to leakage and absorption by adjacent non-fissionable materials, the chain reaction will not occur until a certain mass of uranium is collected that will maintain the action described above; this is known as the *critical mass*. The chain reaction can proceed at a controlled rate with the released neutrons increasing to a certain level and remaining there. When the neutrons produced balance those lost and absorbed, the chain reaction will be maintained; when the total neutron loss and fission absorption exceed the neutron production, the chain reaction ceases. See Figure 2-9.

Breeding

While the uranium isotope 235 is the only *natural* fissionable part (0.7%) of natural uranium, the larger portion, isotope 238 (99.3%) can also play an important part. If an atom of uranium-238 absorbs a stray neutron, it changes (after a delay of several days) into an atom of plutonium-239, an element not found in nature but man made and a fissionable material. The plutonium atoms can replace the uranium-235 atoms consumed in previous fissions. Hence, not only is it possible to utilize more nearly all of the natural uranium as an energy source, but also to manufacture more fuel than is being consumed!

Material, such as uranium-238, which can be converted to fissionable material is called "fertile." Just as fossil fuels are eventually consumed, so too are fissionable and fertile materials. As the nuclear fuel is being continually depleted, it produces "waste products" which act as "poison" by wastefully absorbing neutrons and interfere with the chain reaction. See Figure 2-10. Periodic processing of the fuel removes the accumulation of fission products to be replaced by the addition of fresh fuel.

Most fission products are intensely radioactive, and although many lose their radioactivity very quickly (in minutes or hours) others remain radioactive for years. Fortunately, fission products are formed within the fuel itself and are removed with the fuel to be reprocessed. During reprocessing, the fission products are removed

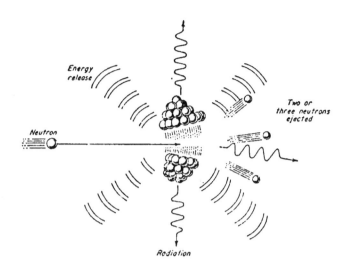

Figure 2-8. A U-235 nucleus hit by neutron fissions (splits) into two smaller atoms, ejects two or three neutrons as well as radiations and energy. *(Courtesy National Industrial Conference)*

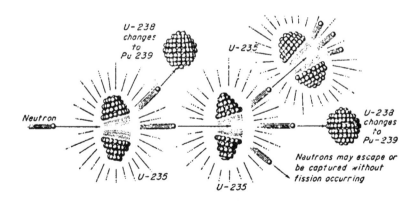

Figure 2-9. In chain reaction, neutrons emitted by fissioning U-235 can fission other U-235 atoms. When a U-238 atom captures a neutron, it most often transmutes to plutonium after several days. Fissioning atoms raise the temperature of the mass of which they are a part. *(Courtesy National Industrial Conference)*

and placed in safe storage in the form of radioactive waste concentrates. The plutonium may also be removed in the process and either reused in place or used elsewhere for power or weapons purposes. See Figure 2-11.

An exceedingly small amount, some one-hundredth to one thousandth of one percent (0.01 to 0.001 percent) of the waste material may be released to the environment (air or water) on a controlled basis in strict compliance with rigorous regulations established and monitored by the U.S. Nuclear Energy Commission.

For technical reasons, the nuclear fuel is replaced when only a fraction of its energy content has been consumed; the reprocessing recovers the unused material.

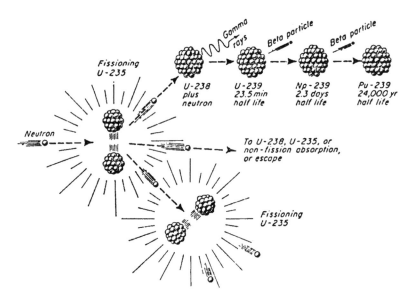

Figure 2-10. Neutrons thrown out by fission fragments may (1) convert fertile U-238 to fissionable Pu-239, (2) fission U-235 nuclei, or (3) be captured by nonfissionable or nonfertile materials, or (4) escape from the reactor entirely. (*Courtesy National Industrial Conference*)

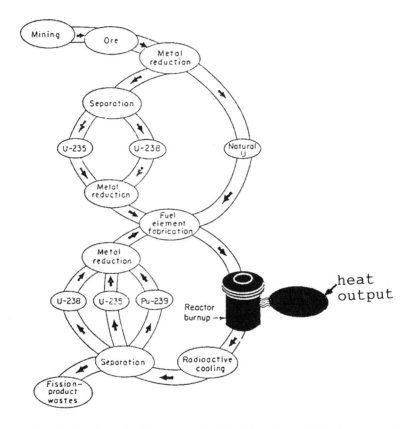

Figure 2-11. Uranium fuel preparation involves the original processing of ore to fuel elements for a reactor. Spent-fuel elements may be reprocessed to recover fissionable material. *(Courtesy National Industrial Conference)*

Chapter 3

Fuel Handling

GENERAL

In almost all cases, the raw materials extracted from mines and wells must be handled and processed before they are suitable for the conversion of their chemical energy into heat. The complexity of the operations and their nature are dependent on the kind and quality of the fuels involved. Only those for coal, oil and natural gas will be discussed here as these are among the principal fuels used in the generation of electricity. The handling of uranium will be included in the discussion of nuclear reactor operations. Other possible fuels listed earlier are subject to special treatment dictated by their nature, availability and local considerations.

DELIVERY

Coal may be delivered to the generating plant sites by railroad, ship, barge, occasionally by truck, and by pipe slurry. The first four methods are rather common and need no further explanation.

In the pipe slurry method of transportation, coal is processed at or near the mine site where it is ground into almost a powder. It is then placed into a pipeline with a great amount of water so that the mixture—called "slurry"—can be handled essentially as a fluid capable of being aided in its journey by pumps. These are strategically placed to keep the slurry moving to its destination. There the water is drained from the slurry and the coal dried for storage or immediate utilization.

Oil may also be transported by rail, ship, barge and pipeline. The oil may vary widely in consistency from a light, easy flowing fluid to a very heavy fluid approaching a tar or asphalt. For oils of grades in the latter thickness, or viscosity, it may be necessary for them to be heated so that they will flow freely as fluids.

Natural gas is usually delivered by pipeline, but may also be transported in liquefied form by railroad or ships.

STORAGE

Coal may be placed in stockpiles or hoppers for long- or short-term storage. Coal is sometimes delivered directly to bunkers for consumption in the immediate future and the bunkers themselves may be included in the plant structure.

Coal Pile

Coal is placed in a stockpile as a reserve against periods when deliveries might be interrupted as well as a supplement to the regular (often daily) delivery during periods, usually seasonal, when demands exceed the normal regularly furnished supply.

Coal piles are usually located outdoors as adjacent to the plant as practical. For protection against annoyance and injury to the public and against pilferage, they are usually fenced in; for esthetic reasons, however, in some cases they may be enclosed by some sort of wall (brick, masonry, plastic, etc.) and may be landscaped as well. See Figure 3-1.

Coal may be piled in horizontal multiple layers so that various sizes are thoroughly mixed in all areas, thus minimizing the absorption or loss of moisture and the movement of air through the pile; each layer is thoroughly packed by successive passes of a tractor.

Good practice often dictates that the top and sides be covered by a thin layer of light road tar or asphalt that may be sprayed upon them. The film essentially seals the coal pile, preventing coal dust being blown into the surrounding atmosphere by wind, creating a nuisance. It also keeps out rain and moisture to a great degree, making less difficult the preparation of the coal for ultimate combustion.

The protective film also inhibits large quantities of air from entering the pile. As the coal settles, air pockets may form which, because of pressures and lack of circulation, can achieve high enough temperatures to effect a spontaneous combustion of the adjacent coal. If a continuous adequate supply of air was available, the coal pile may catch fire, an event that fortunately seldom occurs. The coal pile is inspected regularly with steel rods poked into the pile to locate hot spots and

Figure 3-1. Coal Pile (*Courtesy LILCO*)

collapse the air pockets. The protective film is "patched" or renewed as necessary.

Bunkers

Coal bunkers, installed as part of the plant structure, not only store a given capacity of coal, but function as part of the system in maintaining a continuing supply to the boiler furnaces. Normally the bunker may hold a 24 to 48 hour supply of coal, may take various shapes generally funnel-like with one or more openings at the bottom. They may be built of tile, reinforced concrete, steel plate, with or without acid resisting linings. See Figure 3-2.

The coal bunker should be located so that the flow to the fuel consuming equipment is as nearly vertical as possible. It should also be as far away as possible from flues, hot air ducts, steam pipes or other external sources of heat which might aid in starting bunker fires.

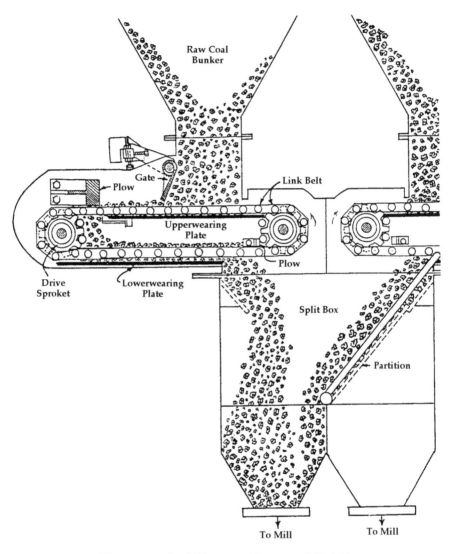

Figure 3-2. Coal Hopper. *(Courtesy LILCO)*

Bunker fires may be started from coal already overheated or burning in storage, or from spontaneous combustion from long storage in the bunker itself. Dead pockets within a bunker should be eliminated.

High moisture, or wet, coal may encounter difficulties in passing through the bunkers. The most frequently encountered difficulties are termed "rat holes" and "coal arches."

Wet coal packs and adheres to the sides of the bunkers to an extent that only the coal directly above the outlet flows out. Often this will progress until a hole slightly larger than the outlet is opened from the top of the pile to the outlet, and the flow of coal may be stopped entirely. This is called "rat holing." When this condition occurs, it is necessary that the hole be filled by mechanical means so that the flow of coal from the bunker be resumed. This is usually accomplished by manually ramming the coal with a long bar or air-lance. The air-lance employs a strong blast of compressed air.

If the wet coal packs solidly over the bunker outlet, a strong arch may be formed that is capable of withstanding the weight of the coal above it. When the coal under the arch is used up, the flow will stop. Air-lances are usually employed to break this arch and allow the flow to resume. Sometimes ports are provided in the base of the bunkers and equipped with compressed air piping that is used to break up such arches.

Coal from the coal pile is brought to the bunkers by lorry, drag-line, conveyor, or other handling equipment.

Oil Tank Farms

Oil is stored in steel tanks located in a "tank farm" adjacent to the power plant. The oil supply is usually stored in a number of tanks so that the loss or destruction of one or more tanks will not adversely affect the operation of the plant. Each of the tanks is erected inside of an earthen or concrete coffer dam or well of sufficient height capable of containing the entire contents of the tank in event of leakage or other failure of the tank. The tanks are also located sufficiently far apart so that a fire in one will not communicate to others adjacent to it. Often, barriers are erected between the tanks as a further precaution; in some cases, the containing coffer dams and concrete wells may be heightened to serve as barriers, or to complement them. See Figure 3-3.

The tanks are usually painted in light pleasant colors not only for esthetic reasons, but to reflect the sun's rays to hold down the temperature of the oil contained within them. The tanks are not filled completely, allowing a cushion of air (or gas) above the oil to permit the expansion and contraction of oil from temperature changes; a safety valve is also provided to relieve the pressure of this air cushion (which may also contain oil vapor) should it rise to undesirable values.

Figure 3-3. Oil Tank Farm *(Courtesy LILCO)*

Since the viscosity of some of the oils precludes their flowing freely, heaters are provided within the tanks to permit their contents to flow out freely.

The arrangement of piping and pumps, both to fill the tanks and to empty them, is such that oil can be moved from one or more tanks to supply the boiler furnaces, to move oil from one tank to another, and to be able to take out of service for maintenance or other reason any of the tanks, the associated pumps, and the sections of piping, without affecting the operation of the generating plant.

Gas Storage Cylinders

Natural gas is brought directly to the generating plant by pipeline and is metered before being utilized. Although the demand may fluctuate, it is not practical to vary the supply of gas to accommodate the changes in demands. For economic reasons, pipelines should operate at as near capacity at all times. To meet fluctuations in demands, therefore, it is desirable to provide storage near at hand to provide for demands beyond the capacity of the pipeline, and to provide for periods when the pipeline may be out of service during emergency, construction or maintenance periods.

In some instances, gas is liquefied under pressure and piped into cylinders where it is stored for reserve purposes. Many of the characteristics of the storage tanks for oil also apply to the cylindrical tanks employed for storage of gas. Gas may also be stored in pipe holders of commercial gas lines pipe laid parallel and interconnected. The pumping and piping arrangements are also somewhat similar. In some cases, the entire installation including tanks, piping, pumps, etc., may be installed underground. See Figure 3-4.

Figure 3-4. Buried Gas Tanks *(Courtesy LILCO)*

Fuel Processing
Coal

From the time coal leaves the bunker until it is burned in the furnace, it is subject to several processes which changes its condition in order that the optimum efficiency be obtained in changing its chemical energy into heat. The selection of the processes to be employed depend largely on the kind of coal available, the size of the generating plant and its vintage, environmental requirements, and on the economics of the situation. See Figure 3-5.

The one process almost always included is the crushing and reduction of the coal to the size necessary. Large modern plants burn coal

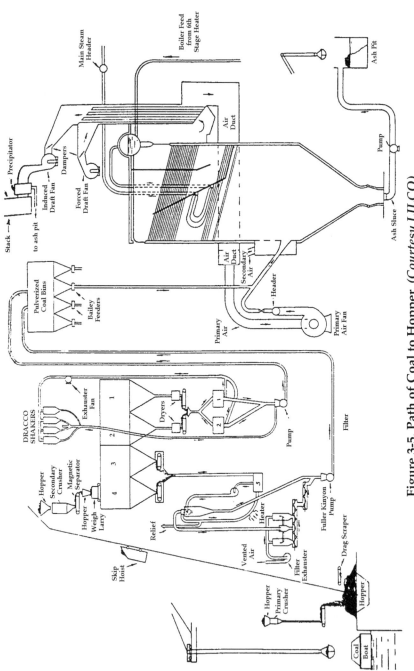

Figure 3-5. Path of Coal to Hopper. *(Courtesy LILCO)*

in pulverized form while smaller older plants generally burn coal in larger sized pieces. While almost all kinds of coal may be pulverized, some of the lower grades, containing large amounts of ash (and trash) may be more easily and more economically burned in larger pieces on grates in the furnace. Coal sizes may be classified as follows:

Table 3-1. Coal Sizes

Classification	Sizes
Broken	3-1/2 to 5 inches
Egg	to 2-1/2"
Stove	to 1-1/2"
Nut	to 1"
Pea	to 1/2"
Buckwheat	to 1/12"

Coal is transported from the storage bunkers to the hoppers associated with the crushing and pulverizing mills by several means: lorries; conveyor belts or drag feeders; gravity chutes or ducts; and other devices.

On the way to the hoppers, the coal may be:

1. Weighed, to determine the efficiency of combustion;

2. Cleaned, by passing over shaking slotted tables or into revolving chambers where gravity or centrifugal force separate out trash and slate; the process may include the coal mixed with water or may be a dry process. Magnetic separators may also be used to remove pieces of ferrous materials (Figure 3-6);

3. Dried, in hot air dryers to increase the mill output as well as boiler efficiency (Figure 3-7);

4. De-dusting, of coal dust or "fines," entrained in the coal, using cyclone separators and bag filters. The fines, if low enough in ash, may be added to the cleaned coal or disposed of separately;

5. Oil spraying, of the moving stream of coal with a film of oil to reduce dustiness. The oil sprayed coal tends to shed moisture and

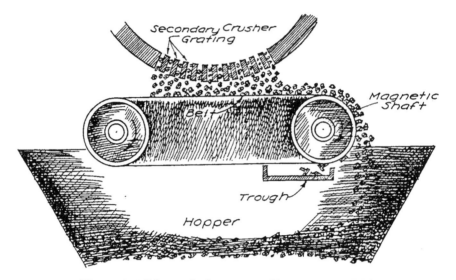

Figure 3-6. Magnetic Separator. *(Courtesy LILCO)*

there is less chance of freezing. The amount of oil used is small, about one to eight quarts per ton of coal.

Screens

Screens may be placed in the path of the coal stream before reaching the crushing mills as well as at the output end of the mills to remove undersized pieces of coal which, in some cases, may be desirable. Screens may be of four basic types:

1. Gravity Bar or Grizzly Type - This consists of a number of sloped parallel bars. The width of the openings between bars, the slope, and the length of the bars determines the separating size.

2. Revolving Type - This type consists of a slowly rotating cylinder set with a slight downward slope parallel to the axis. The envelope of the cylinder is made up of perforated plate or wire screen, the size of the perforations determining the separating size. Because of the repeated tumbling, considerable breaking of coal occurs and, hence, this type screen is limited to sizes of coal smaller than about three inches. Since only a small portion of the screen surface is covered with coal, the capacity per unit area of screen surface is low.

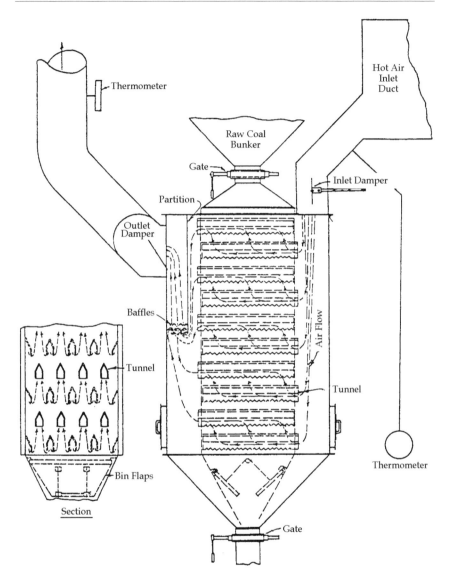

Figure 3-7. Coal Dryer. *(Courtesy LILCO)*

3. Shaker Type - In this type, the screen is mounted in a rectangular frame and may be horizontal—or sloped downward slightly from the feed end to the discharge end. The screen, if horizontal, is given a differential motion to convey the coal along the screen. It is used for sizing and de-watering larger coal sizes.

4. Vibrating Type - The screen is sloped downward from the feed to
    the discharge end, the flow of coal depending on gravity. A high
    frequency low magnitude vibration is given the screen by an elec-
    tric vibrator or by other means. The purpose of the vibrations is to
    keep the meshes clear of wedged particles and to stratify the coal
    so that the fine particles come down in contact with the screen.

Crushing Mills

    Crushing equipment may be of several types; all, however, usu-
ally have some form of screening associated with them. Representative
types are described below:
1. Rotating Cylinder - A large cylinder made up of steel screen plates,
    the size of the screen openings determining the size of the crushed
    coal. The coal fed at one end of the cylinder is picked up by lifting
    shelves and carried up until the angle of the shelf permits the coal
    to drop. Because the gravity force used in breaking the coal is low,
    the production of fines is small. See Figure 3-8a.

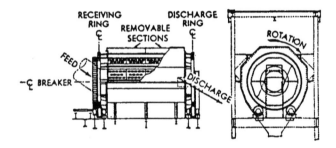

Figure 3-8a. Bradford breaker, for use at mine and plant.

2. Single Roll - A single roll equipped with teeth forces the coal
    against a plate to produce crushing action. Because of the abrasive
    action between the coal and the plate, the quantity of fines pro-
    duced is relatively large. It is commonly used in reducing coal to
    sizes from about 1-1/4 inch (stove) to about 5 inches (broken). See
    Figure 3-8b.

3. Double Roll - The crushing action is obtained by feeding coal be-
    tween two toothed rolls. In rotation the faces of both rolls move

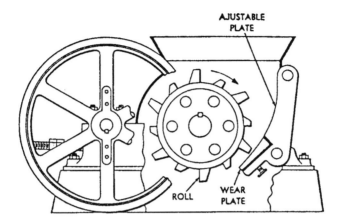

Figure 3-8b. Single roll coal crusher—diagrammatic section.

in a downward direction with the coal. The size of the roll teeth and spacing between the rolls determine the size of the coal. The production of fines, while less than the single roll, is also relatively large. See Figure 3-8c.

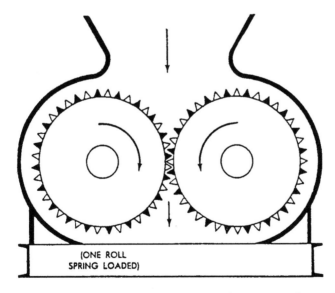

Figure 3-8c. Double roll coal crusher—diagrammatic section.

4. Hammer Mill - This type mill employs the centrifugal force of swinging hammers, striking the coal to crush it. Coal enters from

the top and discharges through grates at the bottom; the spacing of the bars determines the maximum size of the finished product. The direction of rotation may be reversible. Considerable fines are produced which sometimes makes this type mill objectionable. See Figure 3-8d.

5. Ring Crusher - This type is similar to the hammer mill, but employs rings instead of hammers to achieve crushing. See Figure 3-8e.

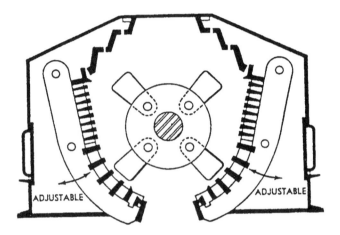

Figure 3-8d. Hammer mill coal crusher—diagrammatic section.

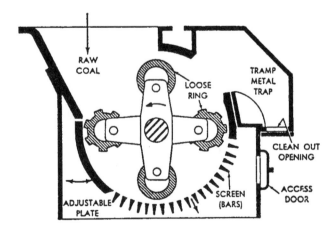

Figure 3-8e. Ring coal crusher—diagrammatic section.

6. Pulverizer - In this type mill, coal is ground to a powder consistency by means of heavy balls. The balls are pushed around a race by means of metal arms keyed to a driving shaft on one end and rounded to fit the ball curvature on the other. As the pushers are rotated, they push the balls before them. Coal enters the top of the mill inside a protecting screen and falls to the ball-races where it is distributed evenly around the balls by intermediate fan blades. As the balls revolve rapidly in the races, the coal is pulverized. There may be two or more tiers of races and balls. When the coal is pulverized to the proper fineness, it is blown upward by fans and thrown against an outer screen by centrifugal force. If the particles are too coarse, they will not pass through the screen but will drop to the balls for further grinding. The protective screen blocks any large particles from tearing the fine screen. The fine particles that pass through the screen are drawn downward by the lower fan blades which, in turn, discharge the coal by centrifugal force to the associated hopper. By increasing the air flow through the mill, it is possible to increase the coal output. See Figure 3-9.

The mills receive coal from other crushers for final grinding. When coal leaves these mills, it has been thoroughly pulverized and is ready to be burned in the furnaces. Fineness is measured by passing the pulverized coal through a series of finely meshed screens. The finer the coal is pulverized, the more effectively it will burn. (A satisfactory operation of the mill will permit about 75% of the pulverized coal to pass through a 200 mesh screen.) Moisture in the pulverized coal increases the difficulty in handling as well as lowers the efficiency of combustion.

As the coal is pulverized, the ground particles are carried away by a stream of air drawn through the mill by exhaust fans. It is necessary to separate the entrained coal from the air stream before passing it on. This is accomplished by "cyclone" type collectors, which utilize centrifugal force to separate the coal and dust from the stream. See Figure 3-10. Coal dust and light particles are drawn off at the top and pass through filter shakers where the coal is filtered through bags and clean air exhausted to the atmosphere. See Figure 3-11. The pulverized coal settles to the bottom by gravity into the hopper from which it will travel to the furnaces.

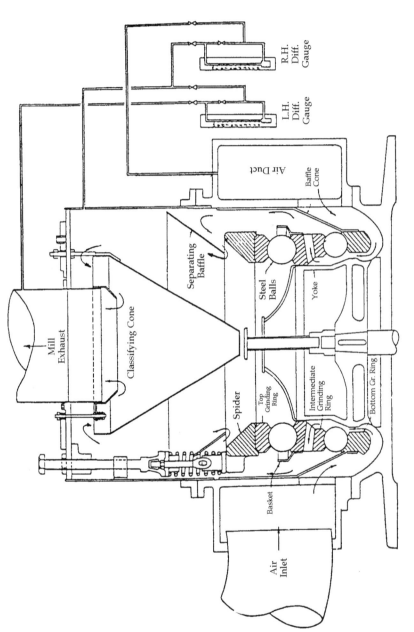

Figure 3-9. Pulverizing Mill. *(Courtesy LILCO)*

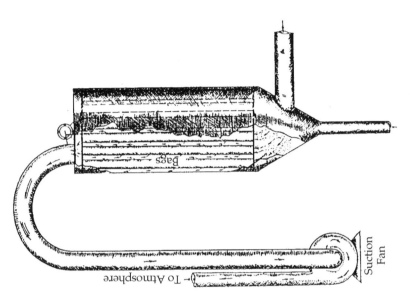

Figure 3-11. Filter Shaker

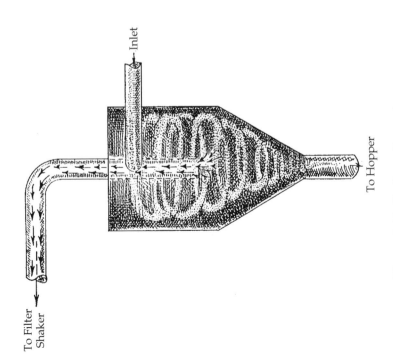

Figure 3-10. Cyclone Separator

Coal Travel

Before the stream of coal is ready to be burned, it is placed into hoppers from which it travels to the boiler furnaces. Dust laden air is exhausted from mills and screens and passed through filter shakers in order to remove the coal dust from the air. The clean air is then exhausted to the atmosphere and the coal dust returned to the system.

The evacuation of dust from dust laden atmosphere is extremely important as the dust combined with a proper amount of oxygen from the air constitutes a potentially explosive mixture.

Chunk coal is transported to the boiler furnace grates by means of chutes and conveyors of several types.

Pulverized coal is transported to the boiler furnace by means of helical (screw) type pumps (See Figure 3-12) which push the coal dust into a piping system where the pulverized coal is aerated and changed from a dense mass to a semi-fluid, in which state it is blown easily through the transport lines to the boiler furnaces. See Figure 3-13.

Fuel Oil

Heated oil is pumped from the storage tanks to the furnaces in insulated piping. It is pumped at a fixed pressure and its volume is measured before it is burned so that the efficiency of combustion may be determined.

The temperature of the oil is critical. It must be hot enough to cause the free and continuous flow of oil, but not so hot as to cause the oil to crack or break down into gases, lighter oils, and other volatile matter.

Screens may be placed in the supply piping to remove any large impurities that may have found their way into the supply lines.

Natural Gas

Natural gas, diluted or enriched, is pumped at a constant pressure from its source to the furnaces in insulated piping to prevent any entrapped moisture from condensing inside the piping. The gas is measured for volume and content of its constituents before it is burned so that the efficiency of combustion may be determined,

Uranium

Uranium-235, in the form of small cylindrical pellets encased in suitable standardized "cladding" is non-radioactive and is delivered to the plant site in ordinary vehicles in perfect safety.

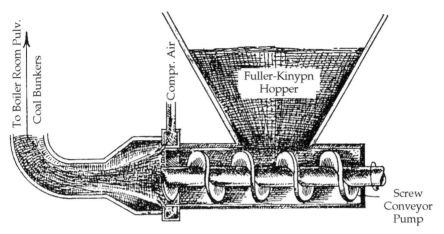

Figure 3-12. Helical Type Pump

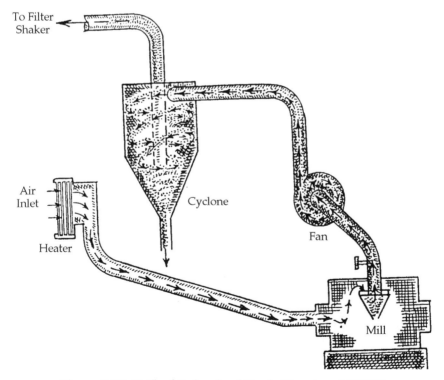

Figure 3-13. Path of Pulverized Coal. *(Courtesy LILCO)*

Chapter 4

Boilers

B oiler is a generic term used to describe a specific type of steam generator. A boiler consists of two main parts: the furnace which converts the chemical energy of the fuel into heat; and the boiler shell or tubes in which the water is converted into steam by means of the heat generated in the furnace.

FURNACE REQUIREMENTS

The furnace requirements include:

1. Should burn completely all of the fuel supplied to it, allowing little or no portion to escape unburned.

2. Should cause this combustion to be as complete as possible by reducing the excess air to a minimum.

3. Should deliver the products of combustion to the boiler proper at as high a temperature and with as little heat loss as possible.

While it may appear a simple matter to meet these requirements, many practical difficulties present themselves in accomplishing this with the highest possible degree of safety, efficiency and reliability, and with the lowest cost of operation. The kind of fuel, the rate at which it is fired, the method of firing, the form and thickness of the fuel bed in the case of coal, the way air is supplied, the size of the furnace and the character of its wall linings, and the method of removing ash and residue, all influence the operation of the furnace, including the most efficient combustion in the furnace.

The optimum design and operation of a furnace largely determine the efficiency of combustion of the fuel used for steam generation. Since there are so many different fuels and grades of fuel, and so many operating conditions, the design of the furnace must meet particular conditions. The aim of furnace design and operation should be to bring the combustible gases into intimate contact with the proper amount of air and to maintain a temperature in the furnace above the ignition temperature of the fuel. Figure 4-1.

Furnace Combustion

The function of the boiler furnace is to convert into heat all the latent chemical energy of the fuel. External heat is applied to the fuel to cause its ignition initially; subsequently the heat is generally supplied by the furnace walls and, in the case of coal, from the bed of glowing fuel.

While combustion is taking place, if the temperature of the elements is lowered, by whatever means, below that of ignition, combustion will become imperfect or cease. Gases developed in a furnace

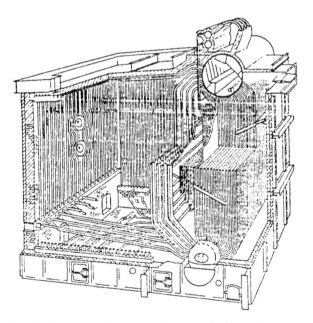

Figure 4-1. Typical aintegral furnace type of boiler. *(Courtesy Combustion Engineering Co.)*

passing too quickly among the tubes of a boiler may be similarly chilled and thus combustion be stopped, causing a waste of fuel and production of large deposits of soot.

Part of the heat developed in the furnace goes from the fuel bed or flame directly into the metal of the tubes by radiation. The rest of the heat raises the temperature of the gases resulting from the combustion—carbon dioxide, nitrogen and water. These gases pass among the tubes transmitting their heat through the tube walls to the water and steam. Thus the gases are cooled and, since they cannot leave the boiler at a lower temperature than that of the water and steam in the tubes, the amount of heat which can be released by the gases is directly dependent on the temperature of the gases when they enter among the tubes. It is important, therefore, that the gases be raised to as high a temperature as possible in the furnace. Hence, every factor affecting this temperature should be considered carefully.

The maximum temperature attained depends on compromises:

1. Excess air is required to achieve complete combustion of the fuel, but as more excess air is supplied, the temperature tends to decrease; if the amount of excess air is decreased to too low a point, the amount of heat liberated will be decreased since incomplete combustion results.

2. So much heat can be generated even with the lowest possible excess air that the temperatures reached may breakdown the enclosing refractory brick of the furnace. The absorption of heat by the enclosing brick together with a large area of water-cooled surface exposed to the heat, may lower the temperature of the furnace and result in poor efficiency.

3. Since the quantity of heat radiated from a burning fuel is dependent upon the duration as well as the temperature, the temperature of the fire will increase as the rate of combustion increases (the relation of fuel to air remaining constant). Rates of combustion of fuel must be matched by appropriate amounts of excess air so as not to produce excessively high temperatures that may cause rapid deterioration of the refractory brick of the furnace.

To protect the brickwork, temperatures are held down by water screens surrounding the furnace, or by water circulated in piping behind or within the brickwork. This permits higher temperatures of combustion to exist with greater heat absorption by the boiler heating surfaces (carrying the water to be turned into steam) both by radiation and by convection from the hotter gases emanating from the fire. The result is greater boiler efficiency.

Furnace Volume

The furnace portion of the boiler and its cubical volume include that portion between the heat sources (grates for coal, jets for pulverized coal, oil and gas) and the first place of entry into or between the first bank of boiler tubes.

The most suitable furnace volume of a boiler is largely influenced by the following:

1. Kind of fuel;
2. Rate of combustion;
3. Excess air; and,
4. Method of air admission.

Kind of Fuel

Anthracite coal needs no special provision for a combustion volume until the boiler is forced to high ratings. The fixed carbon, which comprises a large percentage of the coal, is burned near or on the grate. The carbon monoxide gas rising from the fuel bed requires some combustion space to mix with the air thoroughly, but not much volume in comparison with the highly volatile ingredients found in oil and gas.

Bituminous coal is high in volatile content and requires considerable furnace volume because a large portion of the volatile combustibles must be burned above the fuel bed. In burning high volatile coal, the distillation of the volatile matter takes place at a comparatively low temperature. This condition is favorable to the light hydrocarbons which are more readily oxidized than the heavier compounds which distill off at higher temperatures. Slow and gradual heating of the coal is necessary to bring about the desired results. Once the volatile matter starts distilling off, it must be completely oxidized by the proper amount of air in order to approach smokeless combustion.

In burning pulverized coal, oil or gaseous substances, the important factors are the volume of fuel to be burned, the length of the flame travel, and turbulence.

In general, the greater the percentage of volatile matter present in fuel, the larger must be the combustion space but these two factors are not directly proportional.

Rate of Combustion

When a boiler is operating at a low rating, the fuel and air have quite a period in which to mix and burn completely. As the rating is increased, both the fuel and air are increased. A proportion of the uniting of oxygen and carbon monoxide takes place in the furnace chamber. If the boiler rating is increased to such an extent that the mixing has too short a duration in the furnace volume, the gases will enter the tube area and ignite, producing what is known as secondary combustion, unless they have been cooled below ignition temperatures by the exchange to the tubes. The higher rating expected from a boiler, therefore, the larger should be the combustion space.

Excess Air

To produce efficient combustion each certain grade of fuel requires a definite amount of air to unite with a pound of the fuel. The amount of air varies according to the ingredients of the fuel. The furnace volume must be such that the air required has sufficient time to unite with the fuel as well as to take care of the expanded gas at the furnace temperature.

Method of Admission

The method of air admission is dependent upon the air required which, in turn, depends upon the kind of fuel. The furnace volume for pulverized coal, oil or gas is large in comparison with a stoker installation not only to admit additional air properly but to provide for the long flames. The different ducts required for air admission necessarily have an effect on the shape of the furnace volume.

Furnace Walls

The walls of a furnace are made of fire resistant materials, usually in the form of bricks or blocks. The material and the bonding methods must be such as to withstand extreme and relatively rapid changes in temperature. As the bricks or blocks are heated, sometimes to temperatures approaching 3000°F, they expand. When the boiler is taken out of service for any reason, the temperature of the wall may be in the nature of 70°F, with resultant shrinkage of the bricks or blocks. These changes, unless taking place over a long period of time will tend to cause cracks and failure of the wall. On rare occasions, under extremely high temperatures, the material may fuse and start to deform. Water cooled walls have greatly lessened these harmful effects and reduced maintenance outages and costs. See Figure 4-2.

STOKERS

The combustion of chunk coal requires the furnace to have a grate upon which the burning bed of coals can rest. The grate should provide for the access of sufficient air supply to accomplish complete combustion while maintaining as uniform a bed of coal as possible; it should also provide for the removal of the resulting ash and for the supply of fresh coal.

The fuel bed should be of uniform thickness and density so that the air and gases passing through will meet with equal resistance at all points and will burn the fuel in the same degree, allowing practically no carbon monoxide to pass and only a small excess of air.

To accomplish these chores manually is, in most instances, impractical and mechanical stokers are designed to perform them.

Classification

Mechanical stokers are divided into two general classes, the overfeed type and the underfeed type. In the overfeed stoker, the fuel is fed on to grates above the point of air admission; fuel is fed at a level below the point of air admission in the underfeed type. Commercially, stokers are classified as:

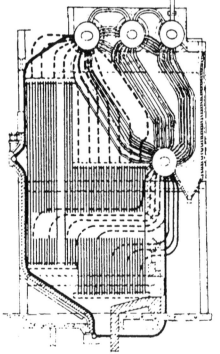

Figure 4-2. Waterwall with top and bottom headers

1. Chain grate or traveling grate type
2. Spreader or overfeed type
3. Underfeed type

Chain Grate or Traveling Grate Stoker

The chain or traveling grate stoker consists of a series of assembled links, grates or keys, joined together in an endless belt arrangement which passes over sprockets or drums located at the front and rear of the furnace. Coal is fed on to the moving assembly and enters the furnace after passing under an adjustable gate which regulates the thickness of the fuel bed. The layer of coal on the grate as it enters the furnace is heated by radiation from the furnace and is ignited with hydrocarbon and other combustible gases driven off by distillation. The fuel bed continues to burn as it moves along, and as combustion progresses the bed becomes correspondingly thinner; so that, at the far end of the travel, nothing remains but ash which is discharged over the end of the grate into the ash pit as the components of the grate pass over the rear sprocket or drum. (See Figure 4-3).

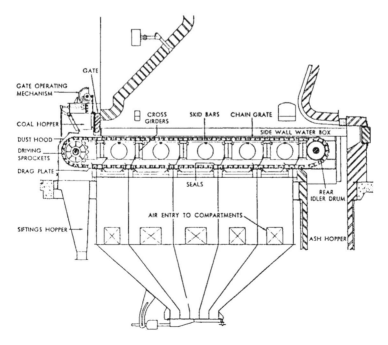

Figure 4-3. Chain grate stoker. Section through a typical installation.

Spreader or Overfeed Stoker

In this type stoker, coal is fed to a stationary grate which may be horizontal or sloped toward an ash pit. Coal is fed on to the horizontal grate by means of a pushing mechanism, and by gravity feed on to the sloped one; the supply of coal is so timed as to insure its complete combustion. The grates are made up of reciprocating or undulating plates which serve to advance the coal from the delivery point to the ash discharge pits.

This type stoker is well adapted for coals with a wide range of burning characteristics, including lignite and some varieties of wood fuel; in general, however, it is not satisfactory for use with anthracite coal. (See Figure 4-4.).

Figure 4-4. Spreader stoker fitted with un- dulating grates.

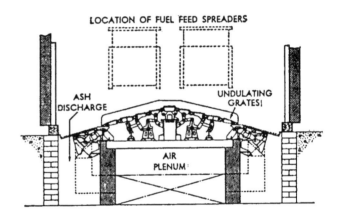

Underfeed Stoker

In the underfeed stoker, coal is force-fed to the underside of the fuel bed, intermittently in small increments by a ram or continuously by a screw or helix. The coal moves along in a longitudinal channel (known as a retort), and as the volume increases, it fills this space and spills over on each side to make and to feed the fuel bed. Air is supplied through vents in the grate sections adjoining the retort on each side (these are known as "tuyeres" with openings shaped so as to drive the air in wide directions and into the fuel bed as far as possible in order to mix the air and coal thoroughly). Figure 4-5.

As the coal rises from the retort it picks up oxygen from the incoming air and heat from the burning fuel above. Gases driven off by distillation burn on their way up through the burning fuel. As the feed moves up into the active burning area it is ignited by conduction

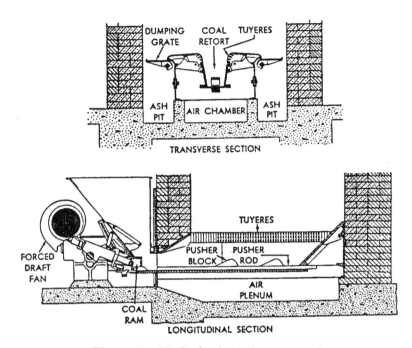

Figure 4-5. Underfeed single retort stoker

from adjacent burning coal. The pressure exerted by the incoming raw fuel and the agitation of the grates cause the burning coal to progress slowly toward the ash discharge or dumping grates. The ash may be discharged from the pit periodically or continuously.

Underfeed stokers may be classified into two types: horizontal feed and gravity feed. In the horizontal feed type, as the name implies, the line of travel of the coal in the retort within the furnace is parallel to the station floor. In the gravity type, it is inclined at an angle, usually 20° to 25°. This latter type consequently requires a basement or tunnel under the floor for ash disposal, while the horizontal feed type only a shallow depression or pit is necessary.

BURNERS (JETS) - PULVERIZED COAL, OIL, GAS

The purpose of burners is to mix the fuel and air in correct proportions and to inject the mixture into the furnace in such a manner that complete combustion occurs in the furnace without injury to the furnace. In practically all burner systems, the entering fuel is already mixed with a certain amount of air (known as primary air) in the feed pipe. The rest of the air (known as secondary air) is admitted through separate openings which lead either directly into the combustion chamber or else into the burner.

Air for combustion must be varied in accordance with the fuel feed (which is determined by the load). But the air carrying the fuel (primary air) must be kept substantially constant as the fuel and air must issue at a speed that will keep the fuel in suspension (about 50 feet per second). Hence, only the secondary air need be varied with the load. Likewise, ignition depends on the proportion in which fuel and air are mixed; each fuel has a mixture ratio at which ignition and the rate of flame travel is at its best. The primary air usually varies from 10% to 30% of the total air required.

The burners consist of jets located either in the walls of the furnace or between the tubes of the boiler. The jets may be so placed that the flame may be fired vertically (Figure 4-6), horizontally (Figure 4-7), or tangentially, as shown in Figure 4-8. The jets in the first two instances may be placed in a fantail fashion resulting in the spreading of the fuel into a broad thin stream. Turbulence is obtained by directing the secondary air across the fuel-laden primary air. Fuel is ejected from the jet by air pressure with the orifices aiding its atomization. In some oil burners, steam is employed in atomizing the fuel. Figures 4-9a and 4-9b.

The most effective turbulence is obtained by placing the burners in the four corners of the furnace, horizontally, in such a manner that the resulting flames impinge on one another, as shown in Figure 4-10. The resulting flame pattern is almost cyclonic in appearance and effect, causing intense turbulence and high temperatures, with stable ignition close to the burners. The flame is relatively short, clear and intense, and the heat liberation is very high. The burners used are of the simplest design and are diverted out some distance from the furnace center to give the desired whirling action. Because of this intense turbulence and high temperatures, no form of refractory covering on the water wall tubes is necessary.

ASH REMOVAL

Ash (and heat) are the products of the combustion of all fuels. The kind and amounts of ash differ with the different fuels.

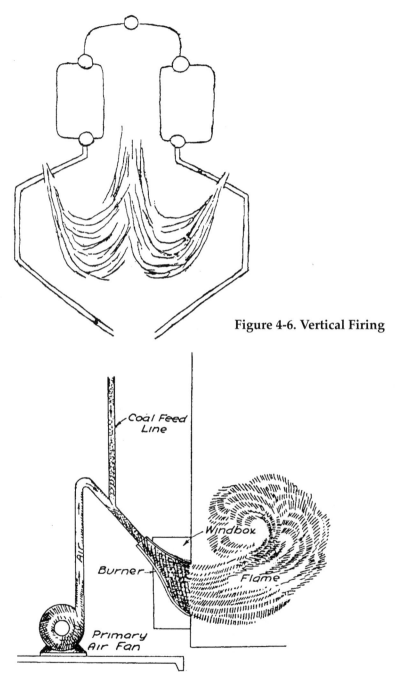

Figure 4-6. Vertical Firing

Figure 4-7. Horizontal Firing

**Figure 4-8a.
Tangential Firing**

Figure 4-8b. Tangential Firing

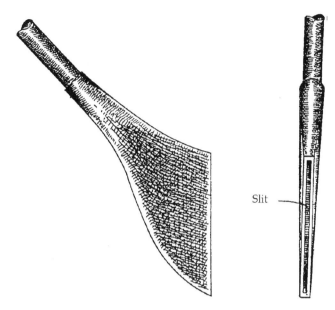

Figure 4-9a. Pulverized Coal Jet Burner

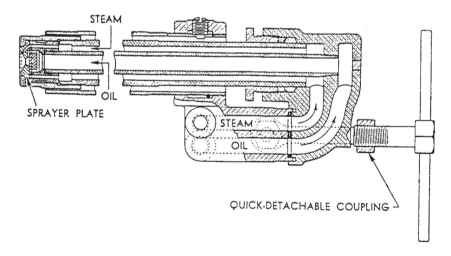

Burner Cross Section. Steam passes through the outer annular passage and the oil through the central tube to a common sprayer plate, mixing after they are discharged from the atomizer.

Figure 4-9b. Steam-Mechanical Atomizing Oil Burner

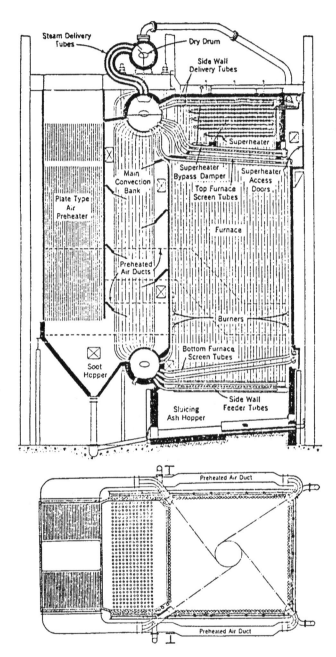

Figure 4-10. Combustion steam generator showing corner-tangential firing.

Ash from burning chunk coal on stokers is collected in ash pits and sluiced out by flows of water. Cinders, clinkers, and slag are removed manually, including "slicing" bars to pry loose slag from stoker parts and furnace walls.

All fuels also produce soot or fly ash; coal also produces cinders. These are small particles of the products of combustion, but may also include some unburned particles, principally of carbon. These are carried away with the gases that ultimately find their way to the atmosphere in the flue, stack or chimney. Their presence in the flue gas is the cause of the color of the smoke leaving the chimney. They are objectionable from an environmental viewpoint and efforts are made to remove them before they reach the atmosphere. There are three general methods of accomplishing this:

1. Dry Method
    (a): Here the ash laden gas is made to change direction by means of baffles located in its path; as the particles are stopped, they fall and rest on the bottom where they can be collected in a hopper. Figure 4-11a.
    (b): Here the ash laden gas enters tangentially into a cone shaped collector. The gas is forced outward as it moves downward, the separating effect increasing as the diameter reduces, clean gas forms a vortex, as in a whirlwind or cyclone, and passes up and out, while the ash swirls down into a collection chamber. Figure 4-11b.
    (c): Here a multiplicity of horizontal tubes is connected to the inlet duct terminating in the gas discharge chamber. A bullet-shaped, spirally finned core, gives the gas a swirling motion,

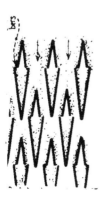

Dust chambers open at bottom to hopper and partly open to upper passage.

Figure 4-11a. Baffles

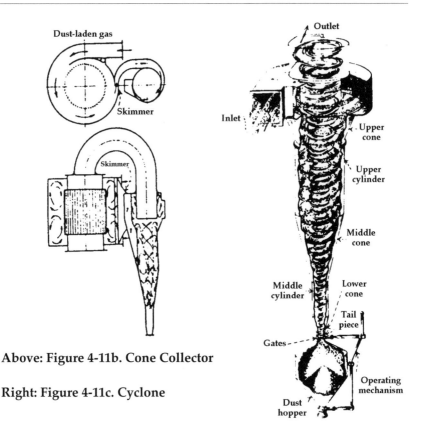

Above: Figure 4-11b. Cone Collector

Right: Figure 4-11c. Cyclone

concentrating the ash in the outer layer of gas. This outer layer of ash laden gas is drawn through an auxiliary cyclone by a fan, from which the clean gas is returned to the inlet duct and ash discharged to the cyclone bottom. Figure 4-11c.

2. Wet Method
    (a): Here the stream of ash laden gas is forced through a water spray; the water sticks to the ash particles, increasing their mass to the point where they are too heavy to be carried in the flue gas and fall to the bottom of the duct, from which they are removed. Figure 4-11d.
    (b): Here the stream of ash laden gas strikes a wetted baffle which turns it at right angles and pass under the baffle and over a body of water. Contact with the baffle and change in direction and velocity causes the ash to be deposited in the water bath, from which it is removed. Figure 4-11e.

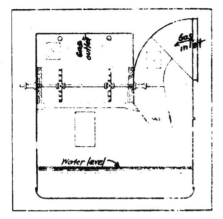

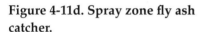

Figure 4-11d. Spray zone fly ash catcher.

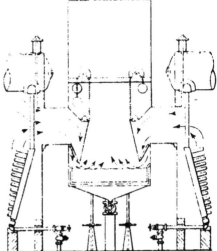

Figure 4-11e. Murray washer an early form of wet cinder catcher.

(c): Here a combination of the baffle and the spray methods are combined in removing the ash from the stream of gases.

3. Electrostatic Method - Here the ash laden gas is passed through a plurality of ducts in which are hung wire discharge electrodes, the duct will forming a collecting electrode. A high electrical voltage, between 50,000 and 100,000 volts DC, is maintained between the electrodes. The walls are grounded but constitute negative electrodes. The ash particles passing through acquire a negative charge and are attracted to and adhere on the positive charged walls. The walls are periodically scraped, allowing the ash to fall, or precipitate, into a collecting hopper. Figure 4-11f.

AIR SUPPLY - DRAFT

Air for combustion is supplied by means of a current, known as "draft." Draft is the difference in pressure available for producing the flow of gases through the furnace and associated stack to the atmosphere.

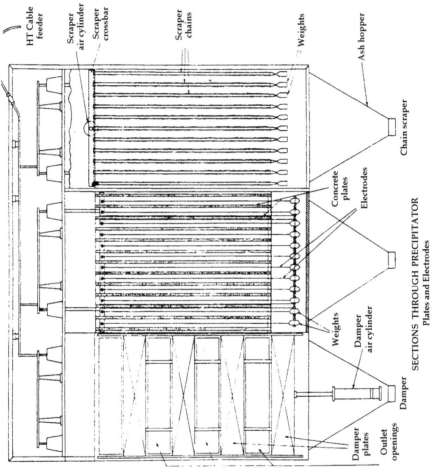

HT Cable feeder

Scraper air cylinder

Scraper crossbar

Scraper chains

Weights

Ash hopper

Chain scraper

Concrete plates

Electrodes

Weights

Damper air cylinder

Damper

Damper plates

Outlet openings

SECTIONS THROUGH PRECIPITATOR
Plates and Electrodes

Figure 4-11f.

Natural Draft

Natural draft is entirely due to the "stack effect" or "chimney effect" produced by the pressure difference between the top and the bottom of a stack which causes the products of combustion to flow up the stack.

If the gases within a stack are heated, each cubic foot will expand and the weight of the expanded gas per cubic foot will be less than that of a cubic foot of the cool air outside the stack. The column of hot gases in the stack is lighter than a column of air of equal dimensions outside. The cool air on its way to the stack passes through the furnace connected to the stack, and it, in turn, becomes heated. This newly heated gas will be displaced by more incoming cool air and the action will be continuous. This causes a difference of pressure which causes the flow of gases into the base of the stack. The hotter the gases and the higher the stack, the greater is this difference in pressure.

The natural draft increases with the height of the stack and the temperature of the gases. Radiation of heat decreases the temperature of the gases from the base to the top of the stack. Surrounding hills, buildings, structures and other obstructions decrease the draft. Variations in atmospheric pressure may affect the draft as much as ten percent; with a high atmospheric pressure and temperature varying from zero degrees in winter to 90+ degrees in summer, the gain in draft may amount to as much as 50 percent.

The necessary height of a stack is determined by the total resistance encountered by gases from the furnace through to the flue; the necessary diameter is determined by the capacity of the associated boiler or boilers. A rough measure calls for 100 feet of stack height for every increment of 4 pounds per square inch of draft pressure, with flue gas temperatures of about 550°F.

Forced Draft

When, for any reason, it is not advisable to make the stack large enough or high enough to supply the required draft to the boiler, resort is had to supplement the available natural draft with air pressure supplied by a fan. Air under pressure from a fan is applied below the grate or the fuel jets in the furnace. The bottom part of the stack will therefore be under a negative pressure or suction while the remainder of the system is under pressure.

If this fan applied pressure is just sufficient to overcome the resistance to the flow of air and gases in the furnace, a "balanced draft" will result. If the forced air pressure is less than sufficient, the stack must furnish the remainder, and the furnace will operate under a partial vacuum, or less than atmospheric pressure. If the forced air pressure is greater than that required, a pressure above atmospheric will exist in the furnace, extending part or all the way through the boiler. This may cause the outward leakage of gases through the furnace-boiler enclosure causing discomfort to personnel and deterioration of the enclosing refractory walls. A forced draft fan may be installed under each individual boiler, or a duct system may be run from one or more fans under a row of boilers. Figure 4-12.

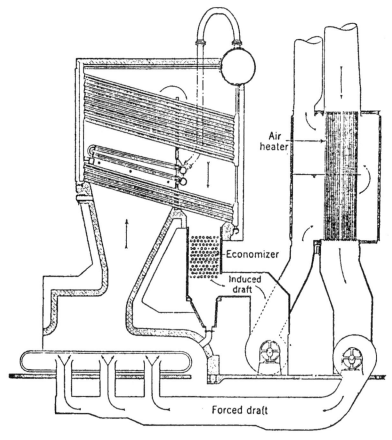

Figure 4-12. Forced- and induced-draft installation.

Induced Draft

If the natural draft of the stack is not sufficient to supply the required draft at the boiler, it will be necessary to install a fan placed so as to draw gases from the boiler and deliver them to the stack; this is known as "induced draft." The induced draft reinforces the natural draft, but tends to cause a greater infiltration of air into the system. Figure 4-12.

Combination of Induced and Forced Draft

Since a forced draft fan tends to produce pressures in the furnace and boiler above atmospheric, and an induced draft fan tends to produce a diminution below atmospheric pressure, a combination of the two should make possible a moving of the gases along their path with a minimum disturbance of pressures. The two can be so regulated to produce a balanced draft sufficient to provide air to carry the load desired while still maintaining a slightly negative pressure in the furnace. Induced draft fans may operate only at times of heavy loads to assist the stack and to prevent the build-up of excessive pressure in the furnace. Figure 4-13.

Fans

There are two essentially different kinds of fans:

1. The centrifugal fan, in which a fluid (air, gas) is accelerated radially outward in a rotor from the heel to the top of the blades and delivered to a surrounding scroll casing. See Figure 4-14a. The blades may be positioned as shown, depending on the velocity required as the fluid leaves the blades.

2. The axial flow fan (fluid accelerated parallel to the fan axis) is not unlike the customary desk fan, but with a casing added for development of a static or constant pressure. See Figure 4-14b.

Since the capacity of a fan is directly proportional to its speed and the pressure produced directly proportional to the square of the speed, control of the air-gas flow system can be managed by varying the speed of the motors operating the fans.

Dampers

Often, it will be found desirable for the fans to operate at varying speeds to accommodate changing requirements for pressure and

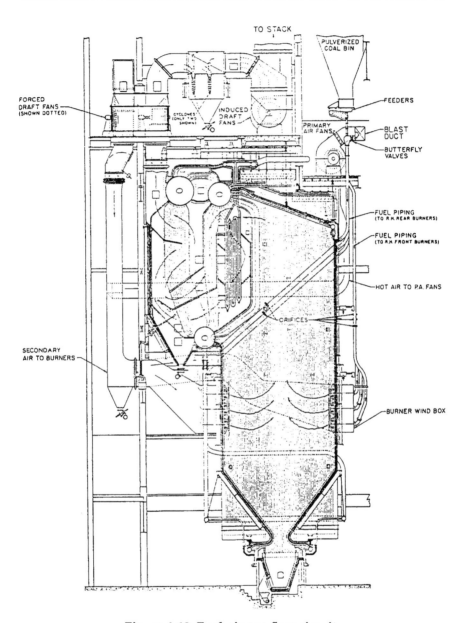

Figure 4-13. Fuel-air-gas flow circuit.

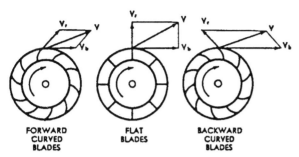

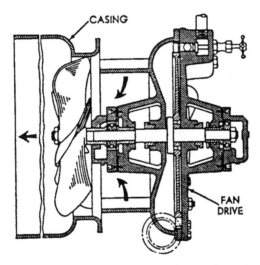

Figure 4-14a. Three general types of centrifugal fans showing relative order of tip speed required to obtain same fluid velocity leaving blades.

V = ABSOLUTE VELOCITY OF AIR LEAVING BLADE
    (SHOWN EQUAL FOR ALL THREE TYPES OF BLADES)
V_r = VELOCITY OF AIR LEAVING BLADE RELATIVE TO BLADE
V_b = VELOCITY OF BLADE TIP

Figure 4-14b. Simple type of axial flow fan.

volume output. If the permissible changes in speed do not meet the requirements of the system, another convenient means of varying the fan output is necessary. A damper placed at the fan discharge, or sometimes in the duct system, furnishes a simple means of control. It acts as a throttle valve, introducing enough resistance into the system to restrict the fan output or flow of air or gases to any desired quantity. Figure 4-14c.

Often, control is achieved by the use of dampers and speed controls on the fan motors. These may be automatically operated by mechanical (hydraulic, compressed air), electrical and electronically actuated devices.

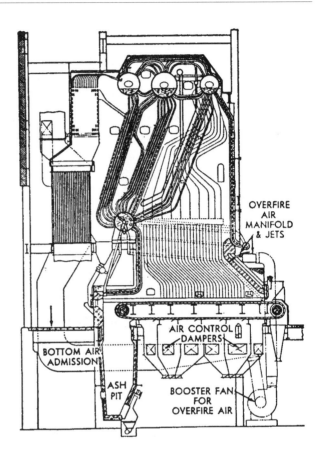

OVERFIRE
AIR
MANIFOLD
& JETS

AIR CONTROL
DAMPERS

BOTTOM AIR
ADMISSION

ASH
PIT

BOOSTER FAN
FOR
OVERFIRE AIR

Figure 4-14c. Air dampers.

BOILER - STEAM GENERATOR

Basic Boiler

Associated with the furnace is the steam generating portion of the boiler; and, although confusing, is also called the boiler.

Boilers are designed to absorb as much of the heat liberated from the fuel as possible. If all the heat liberated in the furnace were absorbed by the boiler, the "boiler" efficiency would be 100 percent. The designs are meant to achieve a high degree of efficiency approaching closely to this ideal. They are designed mechanically for durability to withstand many hours of continuous operation.

There are two types of boilers:

The first is a "fire tube" boiler in which the products of combustion go through tubes and the water surrounds the tubes. This is a low pressure, slow steaming unit and is generally limited in application. Figure 4-15a.

The second is a "water tube" boiler in which water is circulated through the tubes and the gases surround the tubes. This may be either a high pressure or a low pressure unit and has very quick steaming qualities. This type boiler is generally preferable because the amount of water circulated is relatively small, can be quickly heated and rapidly circulated. Water tube boilers are divided into two groups, namely, the straight tube boiler and the bent tube boiler. Figure 4-15b.

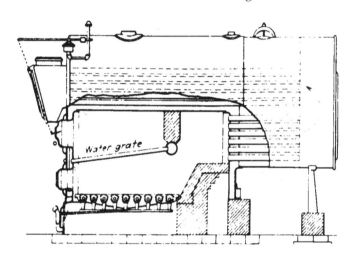

Figure 4-15a. Fire-tube with down-draft grate.

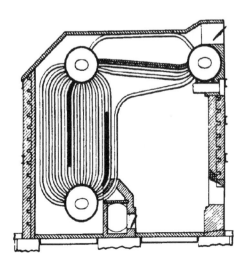

Figure 4-15b. Three-drum bent-tube, L-shape.

Boilers come in a variety of shapes, sizes and complexity. For a basic boiler of the type found in a generating plant, the heating surface consists of two drums connected by circulating tubes. One drum, situated at the lower part of the boiler and called a water drum (sometimes also called a "mud drum") is connected by a large number of seamless steel tubes to the second drum, called a steam drum, situated diagonally at the upper part of the boiler. The tubes, of relatively small diameter (about 4 inches), are grouped in a great number of individual tubes. The ends of each tube are accessible for cleaning or replacing through a hand hole of sufficient size in the wall of the boiler. Figures 4-15c and 4-15d.

Each drum has a front header and a rear header. The front headers of the upper and lower drums are connected together by a number of tubes; similarly, a number of tubes connect the rear headers of the upper and lower drums. The heated water and steam mixture, being lighter than the water, moves from the lower end of the tubes up the incline to the front headers, through which it goes upward to the drum. The steam rises from the water, and the water, in readiness for another cycle, passes in to the rear end of the drum and down the circulating tube to the rear header. The circulation en-

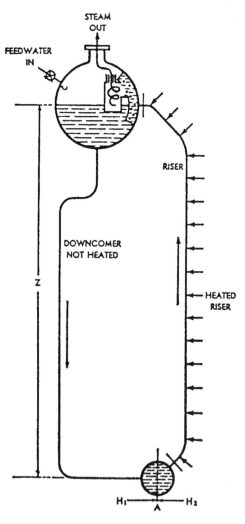

Figure 4-15c. Simple hypothetical circuit (diagrammatic) including primary steam separator in drum.

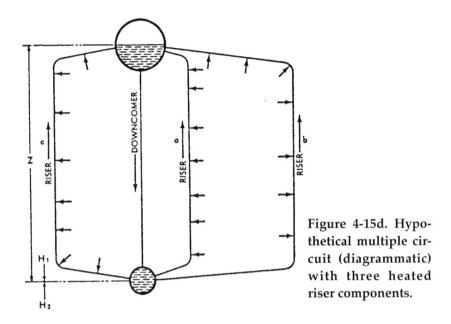

Figure 4-15d. Hypothetical multiple circuit (diagrammatic) with three heated riser components.

ables the water to absorb the heat from the furnace. As the circulation increases, the output of the boiler also increases.

Heat Transfer in Boilers

The heat liberated by the combustion of the fuel in the furnace is immediately absorbed, partly in heating the fresh fuel, but mainly by the gaseous products of combustion, causing a rise in their temperatures.

The heat evolved and contained in the gaseous products of combustion is transferred through the gas filled space and then transmitted through the heating plates or tubes into the boiler water. The process of transmission takes place in three distinct ways. Heat is first imparted to the dry surface of the heating plates or tubes in two ways: first, by radiation from the hot fuel bed, furnace walls, and the flames; and second, by convection from the hot moving gaseous products of combustion. When the heat reaches the dry surface, it passes through the soot, metal, and scale to the wet surface purely by conduction. From the wet surface of the plate or tube, the heat is carried into the boiler water mainly by convection (but also by some conduction). This process is shown pictorially in Figure 4-16, which also portrays graphically the temperature drop from the fire to the boiler water.

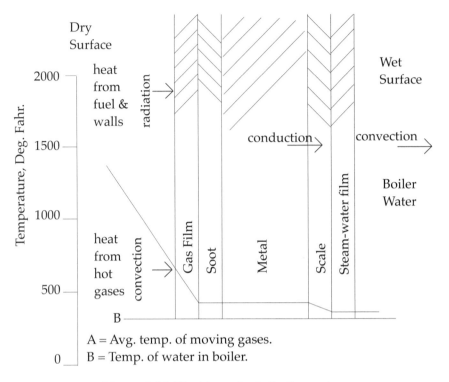

Figure 4-16. Heat transfer to boiler water.

The metal of the plate or pipe is covered with a layer of soot on the gas side and a layer of scale on the water side. In addition, a layer of motionless gas is entrapped in the soot while a layer of water and steam adheres to the scale, or, if the boiler is clean, to the metal.

In practically all boilers, only a small portion of the heating surface is so exposed to radiation from the fuel bed, flames, and furnace walls, as to receive heat both by radiation and convection. By far the greater part of the surface receives heat only by convection from the moving gaseous products of combustion.

The greatest resistance to the flow of heat, that is, the greatest drop in the temperature of the gases, takes place before the hot gases reach the dry surface of the heating metal plate or pipe. If a boiler is even moderately clean, the resistance of the metal itself to the flow of heat (the drop in temperature) through it is very small. The resistance to the passage of heat from the metal into the water (loss of temperature) is also very small. Hence, practically all of the heat imparted to the dry surfaces is transmitted to the boiler water.

Increasing the rate at which heat is imparted to the dry surface of the heating metal plate or pipe increases the rate of steam production in the same proportion. If the initial temperature of the moving gases remains constant, an increase in the velocity with which they pass over the heating metal plate or pipe increases, in an almost direct ratio, the rate at which heat is imparted to the dry surface and, therefore, increase almost directly the rate at which steam is produced.

To increase the capacity of any boiler more gases are passed over its heating surfaces. A boiler that has its heating surfaces so arranged that the gas passage are long and of small cross-section is more efficient than a boiler in which the gas passages are short and of large cross-section.

To increase the efficiency of water tube boilers, the pipes are bent to increase their length and baffles are inserted in such a way that the heating surfaces are arranged in series with reference to the gas flow, thus making the gas passage longer. Figures 4-17a and 4-17b.

Boilers are rated in boiler-horsepower (BHP), one BHP comprising 10 square feet of the boiler heating surfaces; for example, a boiler with 5000 square feet of heating surface would be rated at 500 BHP.

Boiler Efficiency Superheaters

Everyone has observed water boiling in a kettle. With the application of heat, the water becomes hot to the point at which it begins to boil. Further application of heat converts the boiling water to steam, and more heat will increase the presence of steam.

Temperature of Vaporization

At normal atmospheric pressure (14.7 pounds per square inch) the temperature of boiling water is 212°F. If the external pressure on the water is increased, it follows that a greater pressure must be created at the surface (from within the water) before boiling takes place. Hence, more heat must be supplied to the water to bring it up to the boiling point and the water temperature will rise above 212°F before the water will start to boil. Conversely, when the external pressure is less than 14.7 pounds per square inch, less heat is necessary to build up the pressure to the boiling point and, therefore, the water boils at a temperature lower than 212°F. An increase in pressure raises the boiling point of water and a decrease in pressure lowers its boiling point.

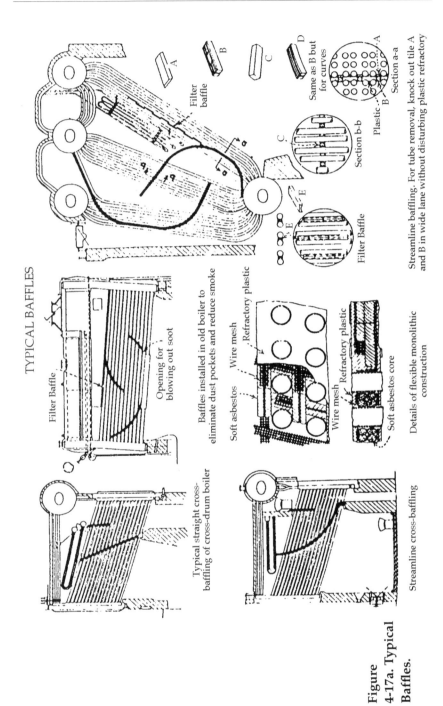

**Figure
4-17a. Typical
Baffles.**

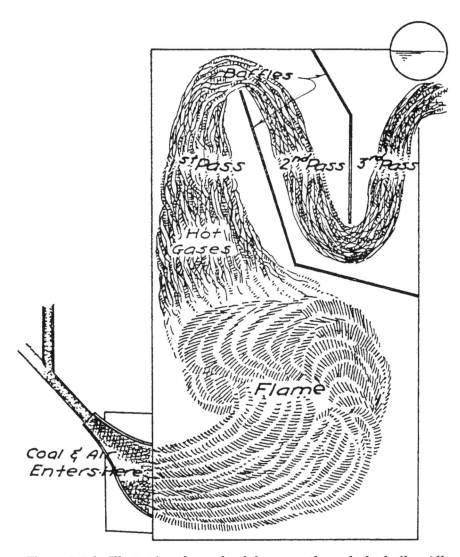

Figure 4-17b. Illustrating the path of the gases through the boiler. All tubes have been omitted.

(Condensation is the opposite of vaporization and it takes place at the same temperature as vaporization if heat is removed from steam. In a condenser the pressure is normally less than atmospheric and condensation takes place at temperatures lower than 212°F.)

Latent Heat

When heat is applied to a liquid causing its temperature to rise, this increase becomes evident to the senses and is sometimes referred to as "sensible" heat. When heat applied to a substance does not cause its temperature to rise, the heat is described as "latent"; for example, the heat required to melt ice or evaporate water.

If heat is added to water at 212°F, the water changes its physical state and turns gradually to vapor, or what is known as steam. This change requires 970.2 Btu for one pound of water at the boiling point changed into steam of the same temperature and at atmospheric pressure. The 970.2 Btu is known as the heat of vaporization for atmospheric conditions. The term "latent heat" is applied to any heat added to a substance when it does not result in a change in temperature.

Saturated Steam

Steam which has been made and withdrawn from the boiler at once, without receiving any more heat, is termed "saturated steam." Theoretically it is nothing but steam and is also known as dry steam. Practically, more or less water in the form of tiny particles thrown up by the violent agitation at the water surface within the boiler accompanies the steam, making it slightly wet. The degree of dryness is usually termed "quality." Thus, if 100 pounds of a mixture passes out of a boiler, 98 pounds in the form of steam and 2 pounds in the form of a mist or spray of water particles, it would be said that the quality is 98 percent.

The steam drawn from the drums of water tube boilers generally contain moisture, perhaps up to 2 percent, when supplied at a higher rate.

To make one pound of dry saturated steam from water at 32°F, therefore, would require an amount of heat equal to a sensible heat of 180 Btu plus a latent heat of vaporization of 970.2 Btu, or a total of 1150.2 Btu.

Specific Volume

When a vessel originally containing water and steam continues to receive heat after all the water has changed to steam, the steam will begin to increase in temperature as soon as the last drop of water has evaporated, and will become superheated. Superheated steam behaves like a gas and is in a far more stable condition than saturated steam. A slight extraction of heat from saturated steam causes condensation, whereas a large amount of heat can be taken (or lost) from superheated steam before liquefaction takes place.

Superheater
    In a boiler, the saturated or slightly wet steam is made to pass from the steam drum to a separate set of tubes inside the boiler where the steam receives more heat than is generated in the furnace. Figure 4-18a. The superheated steam is collected in another drum from which it passes on to the turbine room; in some instances, the drum is omitted and the superheated steam passes directly from the superheater to the turbine room. In the superheater, the temperature of the steam is thereby increased but the pressure remains the same, or drops slightly on account of the friction in the superheater tubes and piping to the turbine room.
    The particular section of a boiler in which the saturated steam from the steam drum is heated to a higher temperature (or to a superheated condition) is known as the superheater. Superheaters are broadly classed according to the source of heat:

1. Convection type—This is usually placed in the gas passages of the boiler where the heat is transmitted by convection. Figure 4-18b, page 106-107.

2. Radiant type—This is situated in the boiler walls which receive radiant heat. Figure 4-18c, page 108.

    The degree of superheat is largely determined by the position of the superheater, the amount of superheating surface, and the velocity of the steam through the superheater tubes. In the case of radiant types, the temperature is affected by furnace temperatures. The radiant type has the advantage that it can be added to existing installations. Figure 4-18d, page 109.
    Often both types of superheaters are used in conjunction with each other as the combination of the two will generally result in a greater utilization of the heat from the burning fuel and in more efficient overall operation of the boiler.

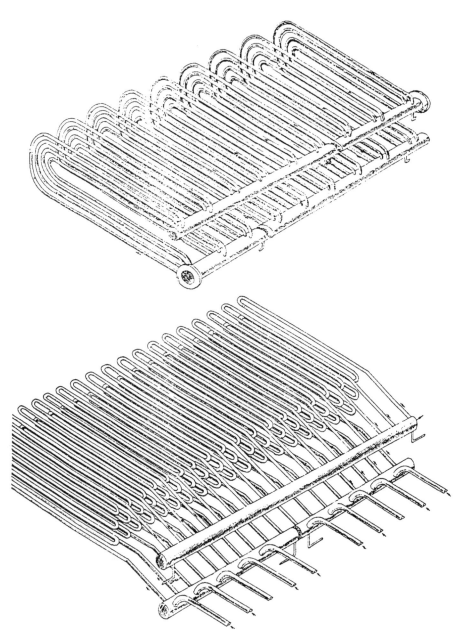

Figure 4-18a. Superheater Tubes.

Since there is no water inside the superheater tubes, they will be subjected to more severe service than the other tubes in a boiler. Special attention is given to the materials used in their design, manufacture and installation.

Steam Tables

The function of a boiler is to receive feedwater and to convert it into steam at some specified pressure and temperature. To calculate the amounts of heat needed for such operations, data regarding the properties of steam are required. It is possible to have any amount of superheat at a given pressure so that in addition to the pressure (which is sufficient in the case of saturated steam) it is necessary to know the temperature to obtain the heat content of the steam. The degrees of heat necessary to raise the temperature of water, evaporate it, and superheat the steam, have been determined experimentally, and the results incorporated in so-called Steam Tables convenient for use in the design, construction, operation and maintenance of the boiler and associated equipment in power plants.

Economizers

Boiler efficiency can be further improved by reducing the flue gas temperature to as near as practical to the temperature of the water entering the boiler.

A feed water heater placed in the path of the flue gas can absorb some of its heat. The hotter the feedwater entering the boiler, the less energy from the fuel burned will be required to raise the temperature of the water to its ultimate value.

The economizer essentially consists of tubes through which the feedwater passes with arrangements provided for the flue gas to pass around the tubes.

Advantages of the economizer are:
1. Less fuel is needed for a given steam flow.
2. Boiler efficiency is higher.
3. Less wear and tear on the boilers because of the high feedwater temperature.

Disadvantages of the economizer are:
1. First cost and maintenance cost.

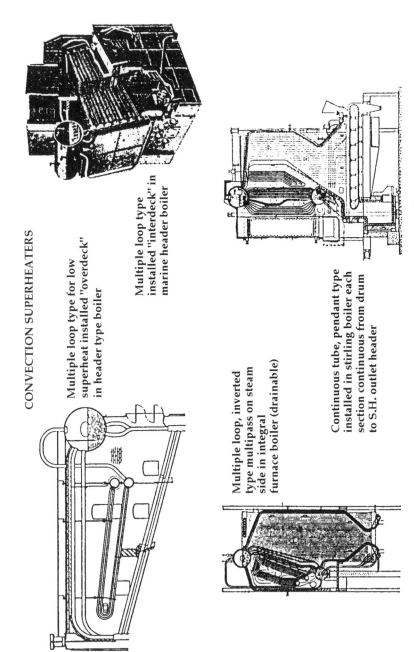

CONVECTION SUPERHEATERS

Multiple loop type for low superheat installed "overdeck" in header type boiler

Multiple loop type installed "interdeck" in marine header boiler

Multiple loop, inverted type multipass on steam side in integral furnace boiler (drainable)

Continuous tube, pendant type installed in stirling boiler each section continuous from drum to S.H. outlet header

Figure 4-18b (Continued). Convection Superheaters.

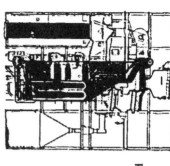

Horizontal multiple loop type
primary, secondary, and
reheater installed in marine
boiler vertical section
through tubes

Continuous tube, pendant type
secondary S.H., and reheater
(nondrainable)
Initial superheater in horizontal
primary section (drainable)

RADIANT SUPERHEATERS

Superheater and reheater in
furnace large open pass boiler
primary superheat, convection
secondary superheat, radiant
reheat, radiant in three stages

Secondary superheater surface
radiant as platens in large cyclone
fired boiler primary superheater and
reheater surface installed in
convection pass

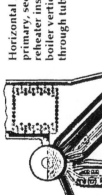

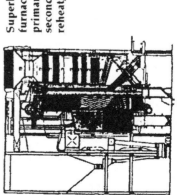

Figure 4-18b&c. Various types and arrangements of convection and radiant superheaters used in boiler units.

TYPICAL SUPERHEATER APPLICATIONS

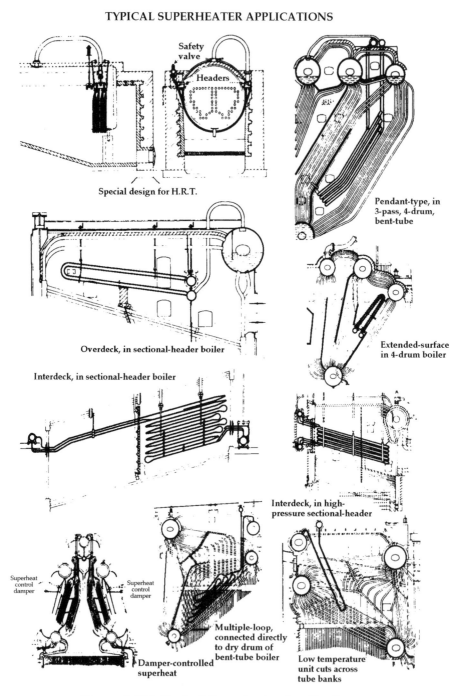

Figure 4-18d. Typical Superheater Applications.

2. More induced draft fan power required.
3. Treatment of water to prevent corrosion.
4. Extra space required for the economizer equipment.

Air Preheaters

The purpose of air preheaters is to use a large percentage of the heat in the flue gas and return this absorbed heat to the furnace in the form of preheated air to the stokers or burners.

Heat is transferred from the flue gas to the incoming air by means of passing the gas through tubes and the air around the tubes.

Air preheaters are often also used to improve the combustion of fuel by drying coal and heating of oil as these fuels enter the furnace.

Feedwater Treatment

Contaminants in the boiler feedwater tend to form: scale that may adhere to the walls of the pipes and tubes; sludge that may clog the pipes and tubes; or both. For efficient operation of the boiler and to reduce maintenance costs, it is desirable to remove or neutralize the contaminants before the feedwater enters the boiler.

Impurities in Natural Waters

Chemically pure water is extremely rare since, being an almost perfect solvent, practically all substances are soluble to some extent in it. Although a great variety of mineral salts or acids may be present in natural water, there are, however, only a relatively small number of these which are to be found in sufficient quantities to be troublesome in boiler feedwater. These generally include salts of calcium, magnesium, sodium, and some other materials that may cause scale deposits and sludge in the metal tubes and piping as well as corrosion of the metal itself.

Suspended solids also may result in the formation of scale forming substances in the boiler. These may be organic or inorganic and consist largely of mud, clay, sewage and other waste products.

Calcium salts may include calcium carbonate (lime, chalk, marble), calcium sulfate (plaster of Paris, gypsum), both of which are very soluble in water. Magnesium carbonate, bicarbonate, sulfate (Epsom salt), chloride, are also very soluble in water. Sodium salts soluble in water include carbonate (soda ash), sulfite, chloride (common table salt) and hydroxide (caustic soda). Silica dioxide, or sand, may combine with some of the other salts and be deposited as scale.

Oils may be of an organic or mineral nature, in suspension or solution; they may cause corrosion, scale deposits, foaming and priming in the boiler.

Free acids from the contamination of water by organic or inorganic substances may result in corrosion of boiler metals.

Gases, carbon dioxide when dissolved in water forms carbonic acid, and with oxygen acting as an oxidizing agent, accelerate corrosion of metal parts. Deaerators are installed to remove these gases. Figure 4-19.

Scale and Sludge

Either sludge, scale or both may be formed from substances or minerals carried in suspension or dissolved in the feedwater. Scale can generally be considered as made up of calcium sulfate, calcium and magnesium carbonates, and silicates. Losses in efficiency due directly to scale are difficult to determine, but may be summed up briefly:

Reduced efficiency due to insulating properties of scale, and economic loss due to boiler outages during cleaning periods. Increased maintenance costs due to cleaning.

Scales, consisting chiefly of calcium sulfate and silicates, are very hard; deposits of calcium carbonate may appear only as a thin porous soft scale that does not build up in thickness.

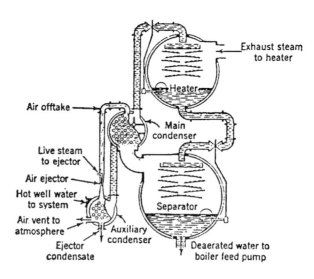

Figure 4-19. Deaerator.

Sludge is formed from precipitates. If some sludges are not removed when the boiler tubes and piping are flushed with clean water (called a "blow down"), due sometimes to sluggish water circulation at points in the boiler, they may settle down on the metal and bake to a hard scale.

Hardness

Hardness is the evidence of presence in water of scale forming matter. It is that quality, the variation of which makes it more difficult to obtain a lather of suds in one water than in another.

Temporarily hard waters are those containing the bicarbonates of calcium which may be precipitated by boiling at 212°F.

Permanently hard waters are those containing minerals such as calcium and magnesium sulfates which cannot be neutralized by heating, except at possibly high temperatures within the boiler itself.

Foaming

Foaming is the production of froth or unbroken bubbles on the surface of the boiler water, and is caused by concentrations of soluble and insoluble salts, together with other impurities such as organic matter which are carried in suspension and thus render difficult the free escape of steam bubbles as they rise to the surface of the water. This condition causes an increase in the moisture content of the steam as it passes to the steam drum.

Priming

Priming is the passing off of steam in belches which may be caused by the presence of too high a concentration of sodium carbonate or sodium chloride in solution.

Alkalinity

A water may be either acid, alkaline, or neutral. The alkaline character of water is brought about by the absorption of impurities of an alkaline nature, as calcium, magnesium, sodium, etc. These impurities are present as bicarbonates, carbonates and hydroxides (that is, in combination with or without carbon dioxide). In steam boilers, the alkalinity (as represented by the bicarbonates) breaks down into carbonates and then into hydroxides. Highly alkaline water has a destructive action on metal causing it to become brittle and weak.

Corrosion

Corrosion or pitting is due to the metal going into solution in the boiler water. Iron "rusts" or combines with oxygen and hydrogen, or with oxygen alone. Unless oxygen or carbon dioxide, or acid forming compounds are present, this process will not take place. The removal of carbon dioxide or other acids by chemical treatment, and the deaeration of the water by preheating will prevent corrosion.

Corrosion is also due to electrolytic action. When two metals of different composition are in contact with each other in water containing impurities in appreciable quantities, an electric battery or cell is formed. All metals in the boiler should have the same electrolytic characteristics; zinc shells or "sacrificial anodes" are sometimes installed which help to mitigate the erosion of the boiler metal.

Preventive Methods

Methods used to prevent the formation of undesirable sludge and scale in a boiler include:

1. Settling tanks together with filtration and water flushing for sediment, sand, mud, organic matter, etc.

2. Settling tanks and chemical coagulants: alum, aluminum sulfate, green vitriol, etc., are used which combine with the sediment in suspension to form a jelly or sponge-like precipitate which can be filtered out.

3. Evaporation or the distillation of water for the removal of all forms of impurities. Figure 4-20.

4. Chemical treatment for removal of substances in true solution in raw water For example, the use of slaked lime, soda, phosphates, and patented commercial preparations, for the removal of calcium and magnesium salts.

While the treatment of feedwater may appear complex and expensive, it is an economic alternative to the reduction in boiler efficiency resulting from clogged tubes and pipes, the costly replacement of boiler parts, and the extensive shutdown periods necessary for remedying the deleterious effects of raw water as feedwater.

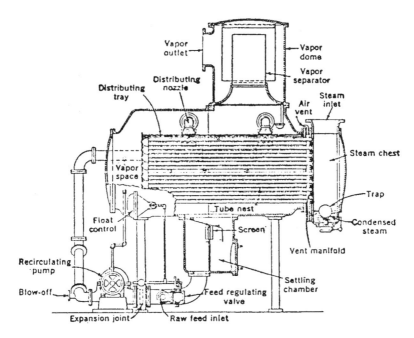

Figure 4-20. Film-type evaporator.

Overall Heat Efficiency

This is shown diagrammatically in Figure 4-21.

Another form of analysis is found in the accompanying heat balance—desirable, but not essential to the study of boiler efficiencies.

Other Efficiencies

Efficiency of installations from an economic viewpoint may be improved by the sale of by-products of the steam producing process. Some of these include:

1. Ash from coal fired boilers for land fill.

2. Fly ash (and soot) from coal and oil fired boilers for road construction.

3. Vanadium from oil fired boilers for hardening of steels.

4. Heat from hot flue gases, steam and steam condensate for commercial heating purposes, including local distribution of steam and hot

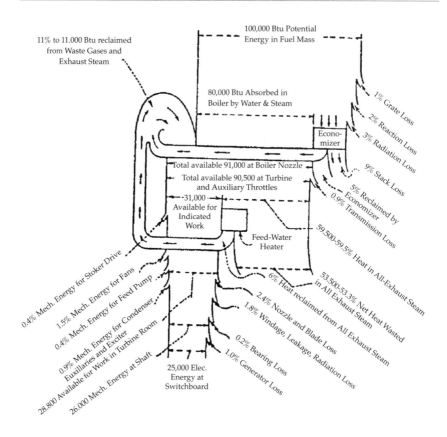

Figure 4-21. Heat stream in an efficient station.
Over-all thermal efficiency = 25%

water, shellfish "farms," and other processes where heat or steam may be employed.

HEAT BALANCE

Efficiency may be defined as the ratio of output to input. In a boiler plant, the input is the total amount of heat in the fuel consumed while the output is represented by the total amount of heat in the steam generated.

The difference between the input and the output in any case represents the various losses, some of which are controllable and some uncontrollable. The heat equivalent of these losses, especially the avoidable ones, is of great importance to the power plant engineer, as a careful study of the variables affecting these losses often leads to clues for reducing them. To this end, a heat balance is of inestimable value, for it provides the engineer with a complete accounting of the heat supplied to the plant and its distribution to the various units.

A complete heat balance is composed of the following items, each of which will be discussed individually:

1. Loss resulting from the evaporation of moisture in the fuel (moisture loss, fuel).

2. Loss from the evaporation of moisture in the air supplied for combustion (moisture loss, air).

3. Loss of the heat carried away by the steam formed by the burning of the hydrogen in the fuel (hydrogen loss).

4. Loss of the heat carried away by the dry flue gases (dry gas loss).

5. Loss from burning to carbon monoxide instead of carbon dioxide (incomplete combustion loss).

6. Loss from unburned combustible in the ash and refuse (combustible loss).

7. Loss due to unconsumed hydrogen, hydrocarbons, radiation, and unaccounted (unaccounted loss).

8. Heat absorbed by the boiler.

9. Total heat in 1 lb. of fuel consumed.

These items are computed on the basis of the number of Btu per lb. of fuel; either, "As Received," "As Fired," "Dry," or "Combustible" being selected as the pound base. Whichever base is selected, all items in the heat balance have the same base. For example, suppose that the "As Fired" base is chosen as a basis for computations, then whenever the words or symbols indicating "per lb. of fuel" appear, the inference "per lb. of fuel AS FIRED" is intended.

Moisture Loss, Fuel (1)

This loss represents the heat necessary to transform the moisture present in the fuel, from water at the temperature of the fuel to super-heated steam at exit flue gas temperature. This transformation takes place in three steps as follow:

(a) To change the water at temperature of fuel, to water at 212°F re-quires (212-T_a) Btu per lb. of moisture. T_a, which is the temperature of fuel, is usually taken to be the same as the temperature of the boiler room.

(b) To change the water at 212°F to saturated steam at 212°F requires 970 Btu per lb. of moisture.

(c) To change the saturated steam at 212°F to superheated steam at exit flue gas temperature requires 0.46 (t_e–212) Btu per lb. of steam; T_e being the exit flue gas temperature in deg. Fahr.

If "M" represents the weight of moisture contained in 1 lb. of fuel, the total loss in Btu per lb. of fuel is then

$$= M \left[(212 - T_a) + (970) + 0.46 (T_e - 212) \right]$$

Moisture Loss, Air (2)

To compute this item, the weight of dry air supplied for combustion, per lb. of fuel from the formulamust first be determined.

$$W_A = 3.032 \cdot \left(\frac{N_2}{CO_2 + CO} \right) \cdot C_b{}^*$$

where N_2, CO_2 and CO represent the percentages by volume of nitrogen, carbon dioxide, and carbon monoxide in the flue gases. C, on the other hand, represents the lb. of carbon burned per lb. of fuel which actually appears in the products of combustion.

From the wet and dry bulb thermometer readings and the psychrometric tables, "N" which is the weight of moisture per lb. of dry

---

*Refer to Item 4 (Dry gas loss)

air, may be determined.

Inasmuch as this moisture enters the furnace as water vapor and not as liquid water, the loss is represented by only the heat necessary to raise its temperature from initial, or boiler room, temperature to that of the exit flue gases, which, per lb. of moisture (water vapor) is equal to 0.46 $(T_e - T_a)$ Btu.

The total loss chargeable to this item may then be written in terms of Btu per lb. of fuel as

$$= W_A \times N \times 0.46\ (T_e - T_a)$$

This loss is rather small and is often included with the unaccounted loss.

Hydrogen Loss (3)

This loss results from the same causes as Item 1 (Moisture Loss, Fuel); 1 lb. of hydrogen uniting with 8 lb. of oxygen to form 9 lb. of water. On this basis, the loss in Btu per lb. of fuel is

$$9H \times [(212 - T_a) + (970) + (T_e - 212)]$$

H is the lb. of hydrogen per lb. of fuel.

Dry Gas Loss (4)

This loss is represented by the heat remaining in the dry products of combustion upon their discharge from the boiler; the word "boiler" being understood to mean the entire unit, consisting of the boiler proper, economizer and air heatr—if so equipped.

$$\text{lb. dry gases per lb. of fuel} + W_G = \frac{11\ CO_2 + 8\ O_2 + 7(CO + N_2)}{3\ (CO_2 = CO)} \cdot C_b$$

where CO_2, CO and O_2 represent the percentages by volume of carbon dioxide, carbon monoxide, and oxygen in the flue gases. C_b is the weight of carbon actually burned and represents the carbon supplied by the fuel less the unburned carbon in the ash and refuse.

C_b = lb. of carbon per lb. of fuel – lb. of unburned carbon in the ash & refuse

C_b = $C - C_u$

The amount of unburned carbon in the ash and refuse can be measured by a laboratory analysis, and is given in percent combustible per lb. of ash and refuse. The lb. of ash and refuse can be determined by actually weighing the ash and refuse, or if this cannot bedone it can be calculated.

$$\text{lb. of ash and refuse per lb. of fuel} = \frac{\% \text{ Ash}}{100 \pm \% \text{ combustible in ash \& refuse}}$$

% Ash = Amount of ash in uel for ultimate analysis.

Then C_u lb. of ash and $\quad \% \text{ Combustible in ash \& refuse}$
refuse per lb. of fuel $\qquad \qquad 100$

The Btu loss from this source per lb. of fuel is then

$$= W_G \ (T_e - T_a) \times 0.24$$

0.24 is approximately the specific heat of dry flue gas.

As has been previously stated, an increase in the amount of excess air supplied for combustion results in an increase in the flue gas temperature. A decrease in CO_2, this directly affects the "dry gas loss."

Incomplete Combustion Loss (5)
    When 1 lb. of carbon burns to carbon dioxide, it liberates 14,550 Btu. When 1 lb. of carbon burns to carbon monoxide it liberates 4,500 Btu. The loss is then due to the incomplete combustion of 1 lb. of carbon is equal to (14,550 – 4,500) or 10,050 Btu.
    The weight of carbon monoxide in the exit flue gases is represented by the expression

$$\frac{CO}{CO_2 + CO};$$ the symbols bing percentages by volume.

Loss due to incomplete combustion of fuel $\dfrac{CO}{CO_2 + CO} \cdot {}^*C_b \cdot 10.050$

Combustible Loss (6)

This loss is represented by the presence of unburned carbon in the ash and refuse. It if were possible to burn all the carbon in the coal this loss would not exist.

In (4) the percent of combustible (carbon) per pound of ash was determined. This percentage multiplied by 14,550 (Btu per pound of carbon) will give the Btu in one pound of ash and refuse.

Combustible loss per pound of fuel = pound of ash and refuse per pound of fuel × Btu per pound of ash and refuse.

Unaccounted Loss (7)

This item, as its name implies, represents the losses which cannot be accounted for and, as such, is equal to the difference between Item 9 and the sum of Items 1, 2, 3, 4, 5, 6, and 8.

Heat Absorbed by the Boiler (8)

This item represents the output of the boiler and is the amount of heat necessary to transfer feedwater into steam. To compute this item, first calculate the evaporation "E" which is the lb. of steam generated per lb. of fuel. The amount of heat absorbed per lb. of fuel is then equal to:

E × [(Total heat in 1 lb. of steam) − (Feedwater temperature − 32)]

The total heat in 1 lb. of steam may be obtained from a set of standard steam tables; provided the pressure, and the degree of the wetness, or the superheat of the steam is known.

Total Heat in Fuel (9)

This is the Btu in 1 lb. of the fuel, as determined by analysis.

---

*Refer to Item 4 (Dry gas loss).

Table 4-1. Heat Balance Data

Temperature	Degrees Fahrenheit	
Wet bulb thermometer		71
Boiler Room "T_a"		85
Flue Gas "T_e"		324
Feedwater		294
Steam		689
Superheat		238
Steam Pressure - lbs. sq. in. gage		410
Steam Generated - lbs.		1,123,000
Fuel burned - lbs. (as fired)		99,584
Flue gas analysis		
	CO_2 %	16.1
	O_2 %	2.7
	CO %	0
	N_2 %	81.2
Ultimate Analysis - Dry Coal		
	C %	82.85
	H_2 %	4.56
	O_2 %	4.08
	N_2 %	1.38
	S %	1.89
	Ash %	6.24
Moisture in fuel - %		0.993
Btu pe lb. of dry fuel - Btu		14536

$$\text{Refuse percent of dry fuel} \pm \% = \left(\frac{\% \text{ ash}}{100 \pm \% \text{ comb. in refuse}} \right) \qquad 7.9$$

Combustible in refuse - %	21.07
Btu er lb. of rfuse - Btu	3066

$$\text{Btu per lb. of fuel as fired} = 14536 \left(\frac{100 \pm .993}{100} \right) = \qquad 14392$$

$$\text{Total fuel burned} \pm \text{dry} \pm \text{lbs.} = 99584 \left(\frac{100 \pm .993}{100} \right) = \qquad 98595$$

(Continued)

Table 4-1. Heat Balance Data (*Cont'd*)

C_b = carbon burned per pound of dry fuel ± lbs.

$$= .8285 \pm (.079) \cdot \frac{21.07}{100}$$

$$= .8285 \pm .0166 = .8119 \text{ lb. or } 81.18\%$$

W_a = dry air supplied per pound of dry fuel

$$= \frac{3.032 \cdot 81.2}{16.1 + 0} \cdot .8119 = 12.4$$

W_g = dry gas, lbs. per pound of dry fuel.

$$= 11 \frac{(16.1) + 8(2.7) + 7(0 \cdot 81.2)}{3(16.1 + 0)} \cdot .8119 = 12.89$$

N = Weight of moisture in air ± lb. per lb. of dry air

Dry bulb - 85°F
Wet bulb = 71°F   From table - Relative Humidity = 50%
Difference 14°F
Weight of moisture in air at 85°F and 100% humidity   =.02634 lb.
Weight of moisture in air at 85°F and 50% humiity   =.2634 ×.50
                                                        =.01317 lb.

$$E = \begin{array}{l}\text{Evaporation, pounds} \\ \text{of water per lb. of dry coal.}\end{array} = \frac{1123000}{98595} = 11.39$$

H = Total heat per pound of steam (from steam table)
    Steam pressure = 425# abs.)
    Total temp. = 689°F)   H = 1356.9 Btu
    Superheat = 238°F)
H_f = Heat of liquid of feedwater = 294 − 32 = 262 Btu
H_i = Heat absorbed per lb. of steam
H_i = $(H_s − H_f) = (1356.9 − 262) = 1094.9$

(Continued)

Table 4-1. Heat Balance Data (*Cont'd*)

(1)   Heat loss due to moisture in coal - Btu
      =.00993 [(212–85) + (970) + 0.46 (324 – 212)]
      =.00993 (1148.5) = 11 Btu

(2)   Heat loss due to moisture in air - Btu
      = 12.4 ×.01317 ×.46(324 – 85) = 18 Btu

(3)   Heat loss due to water from combustion of hydrogen - Btu
      = 9 ×.0456](212 – 85) + 970 +.46(324 – 212)]
      =.4104 (1148.5) = 471 Btu

(4)   Heat loss due to dry chimney gases - Btu
      12.89 (324 – 85) ×.24 = 739 Btu

(5)   Heat loss due to incomplete combustion of carbon - Btu = 0

(6)   Heat loss due to unconsumed combustible in refuse - Btu
      =.079 × 3066 = 242 Btu

(7)   Heat loss due to unconsumed hydrogen, hydrocarbons,
      radiation, and unaccounted for Btu
      14536 – (11 + 18 + 471 + 739 + 0 + 242 + 12471) = 584 Btu

(8)   Heat absorbed by boiler = 11.39 × (1094.9) = 12471 Btu

(9)   Total heat per pound of dry fuel = 14536 Btu

Table 4-2. Summary

	Heat Balance - Btu	Btu	%
	Btu per lb. of coal (dry) = 14536		
1.	Heat loss due to moisture in coal	11	.08
2.	Heat loss due to moisture in air	18	.12
3.	Heat loss due to water from combustion of hydrogen	471	3.24
4.	Heat loss due to dry chimney gases	739	5.08
5.	Heat loss due to incomplete combustion of carbon	0	0
6.	Heat loss due to unconsumed combustible in refuse	242	1.67
7.	Heat loss due to unconsumed hydrogen hydrocarbons and radiation and unaccounted for	584	4.02
8.	Heat absorbed by boiler	12471	85.79
	Total	14536	100%

NUCLEAR REACTOR

The nuclear reactor takes the place of the boiler (furnace and steam generator) found in fossil fuel generating plants. It is a device for the controlled fission of nuclear fuel. Figure 4-22.

Reactor Parts

A basic power reactor consists of: a core of nuclear fuel; a control system to regulate the number of neutrons present and, hence, the rate of fission; and, a cooling system to carry away the heat that is gener-

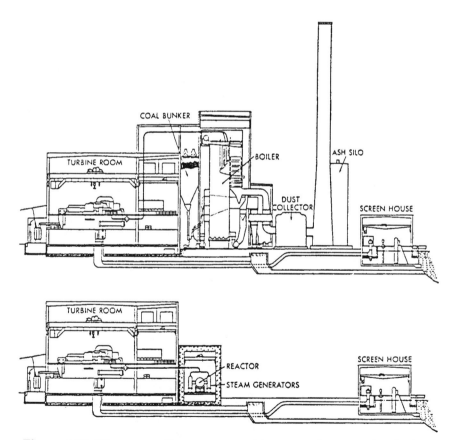

Figure 4-22. Diagrammatic arrangement to approximate scale of a large power plant with conventional (coal) fuel, upper view, and with nuclear fuel for reactor and heat exchanger, lower view. The electrical generating portion of the plant remains essentially unaffected.

ated and keep the fuel from overheating. Figure 4-23. Most reactors also include a moderator, a material that slows down the neutrons to speeds at which they are most effective in inducing fission; often the coolant serves this purpose. Figure 4-24.

In addition, there is a reflector that completely surrounds the reactor core to reflect escaping neutrons back into the core. This reduces the amount of nuclear fuel required as a more efficient use of the released neutrons is made. The moderator sometimes also acts as a reflector.

The entire reactor is surrounded by heavy shielding, usually an enclosure of concrete several feet in thickness for the protection of personnel. The shielding keeps the radiation within the reactor and any neutrons that may stray from the fission process.

Core

The reactor core generally consists of a geometric arrangement of fuel assemblies each consisting of a bundle of metal tubes containing nuclear fissionable material. The material is sometimes diluted with non-fissionable materials designed to slow the neutron flow when the temperature of the fuel rises, and for economy. It is sometimes diluted with fertile material for producing new fuel. The fuel is encased in protective metallic tubes or sheaths (such as zirconium), known as fuel cladding, not only for their desirable mechanical properties, but also as a barrier against the escape of fission products.

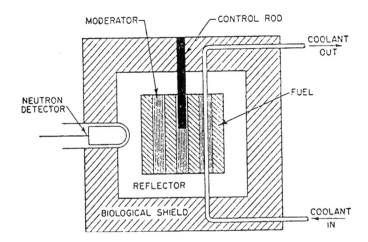

Figure 4-23. Elementary components of a power reactor.

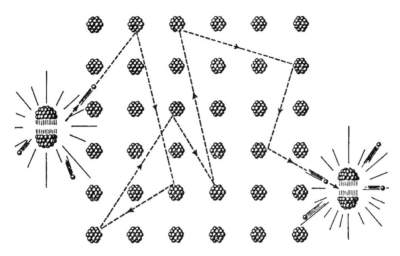

Figure 4-24. Most of the fast neutrons released by fissioning must be slowed to a thermal velocity by bouncing around among the moderator nuclei before they can fission a U-235 nucleus.

The fissionable material consists usually of uranium-235, plutonium-239, or uranium-234, made up into small cylindrical metallic pellets, a number of which are enclosed in the cladding tubes forming the rod. Figure 4-25.

The fuel core is held in place by grid plates in a massively constructed steel vessel.

Control

The reactor is controlled by raising or lowering rods containing neutron absorbing materials (graphite, for example), acting much as an ink blotter. They are moved in and out of the reactor core to soak up some of the neutrons and keep emission of neutrons from becoming excessive. To shut down the reactor, the control rods are fully inserted, thereby depriving the fuel core of the neutrons it requires to maintain the chain reaction to keep the reactor operating. Figures 4-26a and 4-26b.

Cooling System

The cooling system of the reactor carries off the heat from the core to generate steam. Heat must be removed continuously to keep core temperatures from becoming too high, melting it down and destroy-

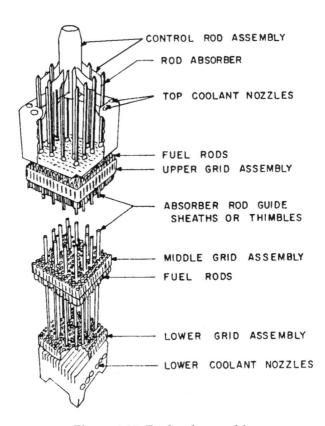

Figure 4-25. Fuel rod assembly.

ing the reactor. Although coolants may include air, helium, hydrogen, fused salts and liquid metals and alloys, water is the cooling medium most generally used. Water enters the reactor vessel and flows upward through the fuel core.

Steam

Steam is formed as the water absorbs heat from the fuel and leaves the vessel at a pressure of about 1000 pounds per square inch. The steam thus produced is radioactive and may be piped to the tur-bine directly. More often the radioactive steam is piped into a heat exchanger which produces non-radioactive steam in separate piping for delivery to the turbine. The same system is used when the coolant may not be water. Figure 4-27.

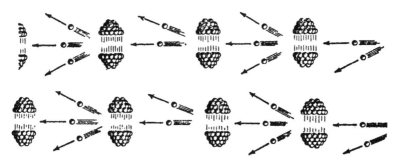

Figure 4-26a. In a steady chain reaction at least one neutron released by fission acts to split another nucleus. Excess neutrons must be absorbed to maintain the number of fissions per unit time at a constant level.

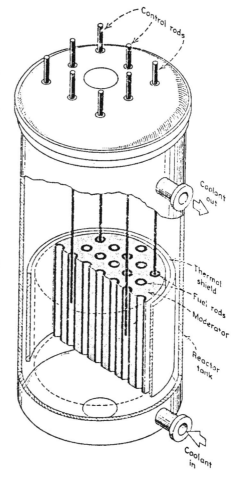

Figure 4-26b. A power reactor might consist of uranium fuel rods held in a moderator surrounded by thermal shielding and supported in a reactor tank. Coolant, flowing upward through the tubes, carries off the heat developed by fissioning uranium atoms. Control rods hold fissioning at the desired rate by absorbing unneeded free neutrons. In some designs, a moderator would be encased in a reflector that bounces the escaping neutrons back into the fuel rods. A biological shield (not shown) completely encases the reactor tank to absorb the lethal gamma radiations.

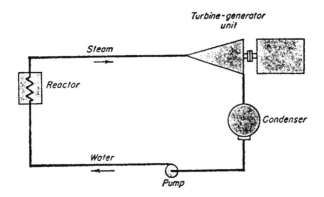

Figure 4-27. Simplest steam-plant arrangement uses a reactor coolant as the working fluid to drive the turbine. This cycle would be the basis of the boiling-water reactor plant.

Safety Measures

The reactor is equipped with various automatic safety devices, many of which are "redundant" (duplicate or more, or back-up) and are designed to "fail safe"—that is, to shut down the reactor in the event the safety devices themselves fail to operate properly.

The reactor housing is designed to contain the consequences of the most serious accident that can reasonably be imagined, and with multiplying factors of safety varying from two upwards. Figures 4-28a and 4-28b.

Nuclear Regulatory Commission (NRC)

The design of the plant, the operating procedures to be followed, and the plant personnel, all must obtain the approval of the Nuclear Regulatory Commission and a number of other national, state and local authorities before permission for construction or a license for operation is granted. The licensing procedure includes a detailed review of the plant's safety not only by the NRC's own regulatory staff, but also by independent experts. And once the plant has been placed in service, its operations are monitored by the NRC.

Contrary to the belief of some people, there is no chance of a nuclear explosion in a reactor core. The nuclear fuel is 30 or more times less concentrated than that required for a reaction in a nuclear weapon.

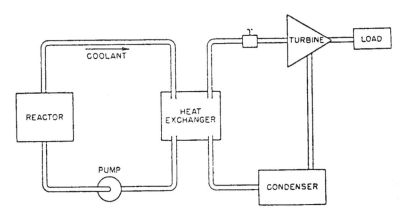

Figure 4-28a. Elementary block diagram of nuclear power plant showing essential elements.

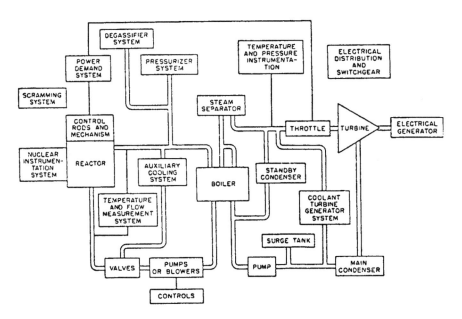

Figure 4-28b. Block diagram of a nuclear power plant indicating some of the important auxiliary systems.

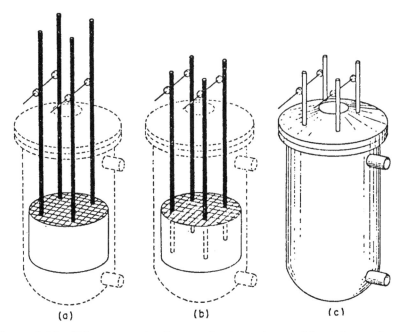

(a)   (b)   (c)

Figure 4-28c. Primary protection against reactor accident depends on inserting scram control rods in (b). The reactor tank (c) must be strong enough to hold pressurized gases generated by the melting core.

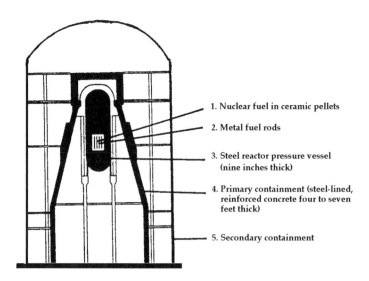

1. Nuclear fuel in ceramic pellets

2. Metal fuel rods

3. Steel reactor pressure vessel (nine inches thick)

4. Primary containment (steel-lined, reinforced concrete four to seven feet thick)

5. Secondary containment

Figure 4-28d. Parts contained within tank.

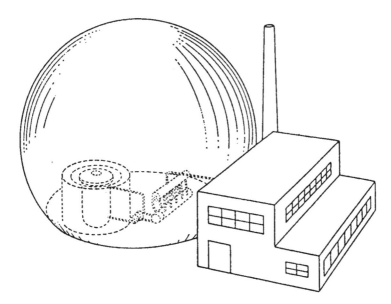

Figure 4-28e. Secondary protection, provided by containment tank or shell, holds the radioactive gases released by the ruptured reactor tank or collant-circuit equipment.

Boiling Water Reactor

In this type reactor, water, which is used as both coolant and moderator, is allowed to boil in the core. The resulting radioactive steam can be used directly but can contaminate the entire steam system and its auxiliaries with radioactivity. Often, however, the radioactive steam is piped to a heat exchanger. Figure 4-29.

Pressurized Water Reactor

In this type reactor, heat is transferred from the core to a heat exchanger by water kept under high pressure to achieve

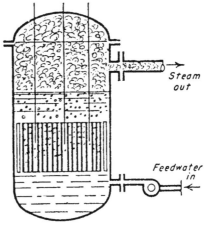

Figure 4-29. In a boiling-water reactor steam bubbles form in the hot reactor core and break through the water surface to fill the upper part of the reactor tank with steam.

high temperatures without boiling in the primary or input part of the heat exchanger. Steam is generated in the "secondary" or output part of the heat exchanger. This type reactor is more prevalent in power plants. Figure 4-30.

Other Type Reactors
Sodium Graphite
　　　This reactor uses graphite as a moderator and liquid sodium as the coolant. This permits higher temperatures to be reached outside of the reactor, but moderate pressures of only about 100 pounds per square inch develop. Because sodium becomes highly radioactive in the reactor, two heat exchangers are employed: in the first, the sodium gives up its heat to another liquid metal, sodium-potassium, which, in turn, carries the heat to the second heat exchanger or boiler. Each coolant loop has its own pump. Figure 4-31.

Fast Breeder
　　　This has no moderator, uses highly enriched fuel in the reactor core and is liquid metal cooled. High speed neutrons fission the uranium-235 in the compact core, and the excess neutrons convert fertile

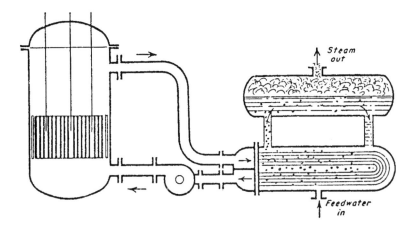

Figure 4-30. In a pressurized-water reactor a pump circulates water between the reactor tank and the heat exchanger of the boiler. Coolant also acts as a moderator in a natural or slightly enriched core. A pressurizing tank (not shown), tapped into the piping circuit, maintains the water pressure at the needed level; the pump acts only as a circulator.

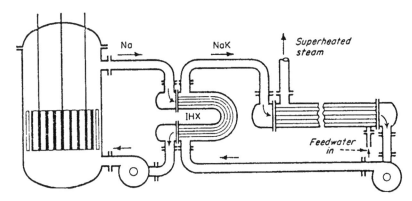

Figure 4-31. Sodium-graphite ractor uses two liquid-metal collant circuits with an intermediate heat exchanger to avoid making steam radioactive in a once-through boiler.

material in the blanket surrounding the core to fissionable isotopes. Like the sodium-graphite reactor, this type reactor system also requires two coolants and two heat exchangers. Figure 4-32.

Homogeneous
This type reactor uses fissionable material carried in a liquid. A pump passes the liquid bearing uranium (in some form) into the reactor, which slows up the flowing uranium to a continuing changing critical mass. Fissioning uranium-235 moderated by water heats the solution which then passes out of the reactor to the heat exchanger boiler to give up its heat. The pump then passes the cooled solution back to the reactor. Figure 4-33.

These three type reactors are more complex in their construction and operation and are generally not employed in electric power plants.

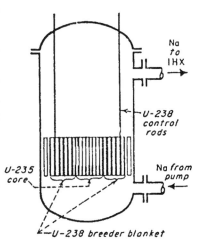

Figure 4-32. Fast-breeder reactor has a U-235 core completely surrounded by a U-238 blanket. The liquid-metal coolant removes heat from both.

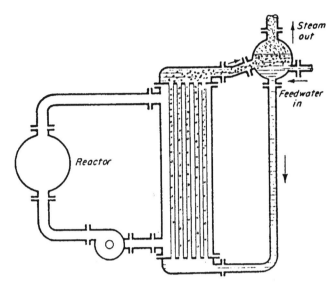

Figure 4-33. A homogeneous reactor has a pump circulating a uranium-bearing fluid between the reactor and the boiler. A critical mass of flowing uranium accumulated in the reactor heats the fluid.

Heat Exchangers

Any device that transfers heat from one fluid, (liquid or gas) to another or to the environment is a heat exchanger. The radiator in a car or those used in space heating are examples of such devices. The heat exchanger is sometimes referred to as the boiler as it accomplishes the same purpose as a boiler in fossil fuel plants. Figure 4-33.

Radioactive Waste

Radioactive waste constitutes equipment and material (from nuclear operations) which are radioactive and for which there is no

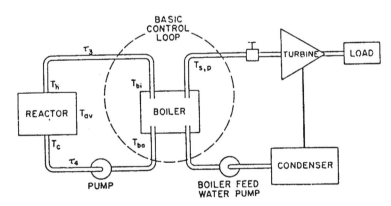

Figure 4-34. Elementary power plant as a basic control loop.

further use. Wastes are generally classified as high level (having concentrations of hundreds of thousands of radiation units per cubic foot), low level (in the range of one one-thousandth (0.001) or micro radiation unit per cubic foot), or intermediate (about midway between these extremes).

The management of radioactive waste material is classified under two categories. The first is the treatment and disposal of very large quantities of materials with low levels of radioactivity. These materials are the low activity gaseous, liquid and solid wastes produced by reactors and other nuclear facilities such as fuel fabrication plants. Figure 4-35a. The second category involves the treatment and storage of a very small volume of wastes with a high level of radioactivity. Figure 4-35b. These high level wastes are by-products from the reprocessing of used fuel elements from nuclear reactors. Unfortunately, these two types of waste are often considered by some as a single entity.

Radioactive wastes produced at nuclear power reactors and other facilities are carefully managed and releases of radioactivity into environments are strictly regulated by the NRC and other governmental bodies.

Reprocessing

Neither the reprocessing of used fuel nor the disposal of high level wastes is conducted at the nuclear power station sites. After the used fuel is removed from the reactor, it is securely packaged and shipped to the reprocessing plant. After reprocessing, the high level wastes are

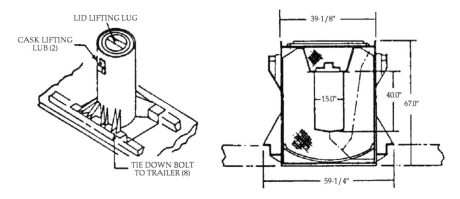

Figure 4-35a. Cask for transporting low-level wastes. (Courtesy Chem-Nuclear Systems, Inc.)

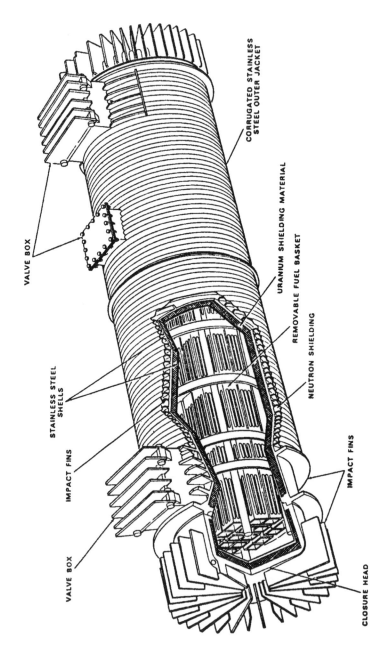

Figure 4-35b. Spent fuel shipping cask, Model IF-300. (Courtesy of General Electric Company.)

concentrated and safely stored in tanks under controlled conditions at the reprocessing plant. As with power plants, the NRC carefully regulates and monitors the operation of such plants.

Storage

Interim storage of high level liquid wastes is safely handled by underground tanks. These liquid wastes are concentrated and ultimately changed into solid form which will then be transferred to a government site, such as an abandoned salt mine, for final storage. These salt mines have a long history of geologic stability, are impervious to water, and are not associated with usable ground water resources.

Treatment and storage systems include short time storage of liquid wastes, evaporation, demineralization, filtration of liquids and gases, and compression of solid wastes. They also include chemical treatments to concentrate radioactive solids and liquids in concrete or other like materials.

Leakage

Quantities of radioactive materials released in power plant effluents have been found to be so small as to be negligible and far below the few percent of authorized release limits. See Figures 4-36a and 4-36b and Table 4-3.

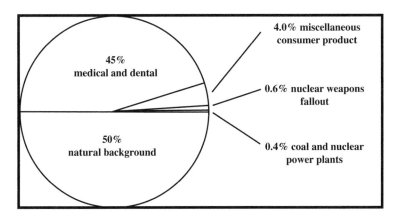

Figure 4-36a. Sources of radiation.

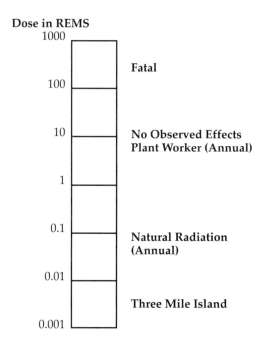

Figure 4-36b. Selected radiation dosages, ranging from the trivial to the fatal.

Nuclear plant operations introduce some slight addition of radiation into the atmosphere. These are well below the limits established as safe by the NRC. (Some background information may be reassuring. In the 1930's, in connection with "radium" painted on watch dials, tests were made to determine the amount of radiation harmful to humans and it was established that humans could safely tolerate an exposure of 291,000 millirems a year. A rem or 'roentgen equivalent man,' is the unit of radiation which produces the same biological effect as a unit of radiation from ordinary X-rays. While the same regulatory bodies determined that 5,000 millirems per year was a permissible limit, it established a limit of 500 millirems as the permissible exposure for the general public. Plant designs further limit this value to about one-thousandth (0.001) or only 5 millirems per year. In contrast, natural background radiation may exceed 300 millirems per year.)

Table 4-3. Radiation Exposure Levels and Their Effects

At this level:	Which is equivalent to:	Over this time period:	The observable effects are:
40 millirem (mrem)	1 chest X-ray	Less than 1 second	None
150 mrem	Average person's exposure	1 year	None
150.04 mrem	Average person's exposure with plant operating	1 year	None
15,000 mrem	Average total exposure	Lifetime	None
25,000 mrem (or less)	625 chest X-rays	Few hours or days	None
More than 25,000 mrem	More than 625 chest X-rays	Few Hours or days	Detectable chromosome changes.
More than 50,000 mrem	More than 1,250 chest X-rays	Few hours or days	Likely temporary changes in blood chemistry; return normal in a few weeks; nausea/vomiting possible.
More than 100,000 mrem	More than 2,500 chest X-rays	Lifetime	Increased frequency of leukemia and some other types of cancer.
More than 150,000 mrem	More than 3,750 chest X-rays	Few Hours or days	Early symptoms: nausea/vomiting/malaise; later: possible infections, fever, hemorrhage, loss of hair, diarrhea, loss of body fluids, effects on central nervous system.
450,000 mrem	22,500 chest X-rays	Few hours or days	50% chance of death.
3,000,000 mrem to parts of body	Cancer therapy	Several weeks	Temporary symptoms of high exposure, but *extension of life also results from treatment of disease.*

Chapter 5

Prime Movers

Prime movers are machines of power plants that convert heat energy into mechanical energy, and which drive the generators that produce electricity. These machines fall into three general categories:

1. Internal combustion engines

2. Reciprocating steam engines

3. Steam turbines Some power plants may contain a combination of these.

INTERNAL COMBUSTION ENGINES

Internal combustion engine plants operate with the combustion of fuel taking place within the prime mover. The fuels commonly employed include natural gas, gasoline, petroleum and its distillates, alcohol, and by-products of industrial and other operations. The fuels used may be in a gaseous state originally; they may be liquid fuels gasified by evaporation; or they may be heavy fuel oils injected into the cylinders of the (Diesel) engines by means of a fuel pump or air jet. The combustion of the fuel may take place within one or both ends of its one or more cylinders.

Although their utilization of the heat energy of the fuel is usually better than that found in steam plants (as the intermediate steps with the heat losses are not necessary), their initial cost, maintenance charges and depreciation restrict their employment to relatively small units for emergency purposes.

Classification

Internal combustion engines may be classified as to:

1. Cycles of combustion: Regular or constant volume; and, heavy oil (Diesel) or constant pressure.

2. Cycles of operation: Two stroke and four stroke cycles.

3. Cylinders: Number (one, two, four, etc.); arrangement (vertical, horizontal, V, single-acting, double-acting).

4. Fuel handling: Vaporizer, carburetor, solid injection, air injection.

5. Cooling: Water, liquid compounds, air.

6. Power output (in horsepower) and speed (in revolutions per minute).

Cycles of Combustion

The combustion of fuel in an internal combustion engine may be accomplished in two ways:

1. The fuel in gaseous or vaporized form is intimately mixed with a suitable quantity of air and the mixture compressed in the cylinder. It is then ignited by an electric spark and the resultant explosion causes a rise in pressure that is practically instantaneous, with a nearly constant volume.

2. Air only is compressed in the cylinder at a very high pressure so that its temperature is sufficient to ignite the fuel without the aid of an electric spark. At the instant this high compression pressure is attained, the fuel is sprayed or injected into the cylinder, at the same time the piston begins to move on its power stroke. The ignition of the fuel produces a further increase in temperature and a practically constant pressure in maintained during this portion of the stroke of the piston; fuel injection can continue and pressure maintained until all the oxygen available is consumed.

Cycles of Operation

This process includes the introduction of fuel and air into the cylinder, their compression and ignition, and the elimination of waste products. This cycle of events may be accomplished in two ways:

1. Four stroke cycle: This is the most usual type of operation and is
    completed in four strokes of the piston or two revolutions of the
    crankshaft. Refer to Figure 5-12 (a) & (b).

    (a) The charge of gaseous fuel and air is drawn into the cylinder,
         the intake valve being open during the downward or suction
         stroke of the piston.

    (b) At the start of the return or compression stroke, both intake
         and exhaust valves are closed and the mixture of fuel and air
         is compressed into the clearance space above the piston (gener-
         ally one-fourth to one-eighth of the original volume).

    (c) Near the end of the upward stroke, the compressed mixture
         is ignited by an electric spark, causing the mixture to ignite,
         resulting in a very high temperature and a drastic rise in pres-
         sure.

    (d) During the next downward stroke, the ignited mixture ex-
         pands, doing work on the piston which, through the connect-
         ing rod and crank, causes the crankshaft to rotate. A small
         quantity of the burned mixture may remain in the cylinder,
         but is mixed with the next charge of fuel.

    After the last cycle, the inlet valve is again opened and the cycle
of operations is completed. In the four-stroke cycle, there is only one
power stroke out of every four.

    With such engines, there is a practical limit to the amount of
compression. The pressure must not be so high as to cause preignition
by unduly raising the temperature of the mixture during the compression stroke, nor so great as to give too sudden an expansion known as
detonation. Detonation is the nearly simultaneous ignition of the mixture
by a compression wave, as distinguished from an explosion in which
ignition speeds from one part to another in a manner like that of slow
combustion. The compression wave heats and ignites the unburned
part of the mixture before the igniting flame reaches it. Detonation may
be distinguished by sharp, metallic, hammer like sounds (pinging and

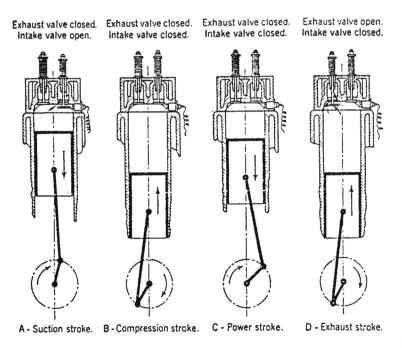

Exhaust valve closed. Exhaust valve closed. Exhaust valve closed. Exhaust valve open.
Intake valve open.   Intake valve closed. Intake valve closed.   Intake valve closed.

A - Suction stroke.   B - Compression stroke.   C - Power stroke.   D - Exhaust stroke.

Figure 5-1a. Cycle of operation in a four-stroke cycle engine.

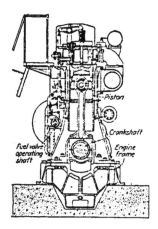

Figure 5-1b. Partial section of 4-cycle diesel engine. (Courtesy Worthington Pump and Machinery Corp.)

knocking). When a small volume of tetraethyl lead (or other compound) is added to ordinary gasoline, the compression pressure may be safely increased. Pressures of from 40 to 200 pounds per square inch may exist, with the lower values present (to about 100 pounds) for gasoline and upper values for natural gas and by-product gases as fuel.

2. Two stroke cycle: Here the operation is accomplished in two strokes of the piston or one revolution of the crankshaft. Refer to Figure 5-2.

   (a) A charge of compressed air is blown into the cylinder, forcing out the exhaust gases through ports that remain open during this part of the cycle.

   (b) The return stroke compresses the air into a small volume attaining high pressures in the clearance space, and causing temperatures to rise greatly.

   (c) Just before the piston reaches its top position, a charge of fuel is injected into the cylinder causing combustion to take place (because of the high temperature) and the piston starts downward continuing to the end of its stroke under the pressure of the expanding gases. Before the completion of the downward stroke, the piston uncovers the ports through which the exhaust gases are forced.

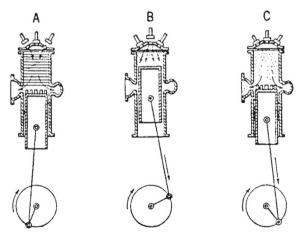

Figure 5-2. Cycle of operation in a two-stroke cycle engine.

The cycle is then repeated, and explosion is obtained every revolution of the engine shaft. Cylinder temperatures may exceed 1000°F and pressures may exceed 500 pounds per square inch.

Cylinders

The number of cylinders is based on two requirements: the relative smoothness of the power supply desired, and the total amount of power to be delivered, modified, of course, by economics and metallurgy.

For the four stroke engine (the most common), there is one power stroke in every two revolutions of the crankshaft. A single one, properly timed, can deliver the same power by two power strokes every two revolutions, with the output approximately half of the unevenness. Continuing this reasoning, six smaller cylinders, each one sixth of the single one, properly timed, can deliver the same power from six power strokes every two revolutions with a resultant six-time improvement or lessened unevenness of rotation. A flywheel, a wheel of relatively great mass will by absorbing energy during the power strokes and because of its inertia of its rotation, continue to turn even during the short interim periods when no power is delivered to the crankshaft dampening out the unevenness, resulting in a smooth and even delivery of power.

The mechanical restraints imposed by stress limits of the metals involved, and the cost differences, usually restrict the number of cylinders and play a part in how they are arranged. Space limitations may also dictate how they may be arranged.

As the demands for power become greater, the cylinders may become larger in dimension, and the combustion of fuel and the explosive expansion of gases may take place on both sides of the piston, in a double acting operation. In some cases, the larger cylinders are limited to approximately two feet in diameter and may require that the piston and connecting rods be cooled; these are made hollow and oil under pressure is used for cooling.

Fuel Handling

In those engines in which air is compressed separately and in which fuel is introduced into the cylinders, the fuel may be vaporized by being passed through a nozzle under pressure or the fuel may be supplied by "solid-injection." In the latter case, liquid fuel is pumped under pressure as a thin stream into the cylinders.

Where air is to be mixed with the fuel before it is introduced into the cylinders, it may be done by means of a carburetor or by "air injection." A carburetor is a device in which the correct air-fuel mixture is formed, at all loads and speeds of the engine, by passing air over or through gasoline or the more volatile fuels. The term is also applied when, in this device, the fuel is atomized or sprayed into the current of incoming air. A simplified diagram of a basic carburetor is shown in Figure 5-3.

In air injection, the fuel is injected with a blast of air at high velocity, and is thus atomized and distributed through the main mass of compressed air in the cylinder. The pressure of the air for the blast must be considerably higher than the highest cylinder compression pressure; the injection pressure may vary from about 700 pounds per square inch for light loads to about 1500 pounds for maximum loads. The compressor for the air is driven directly by the engine and may use as much as 15 percent of the power developed.

Cooling

Except for small gasoline powered engines that are cooled by an air stream flowing past the engine, practically all other internal combustion engines are cooled by water, liquid chemical compounds (that may also be rust preventatives), or a combination of the two. The coolant is circulated by pump through jackets surrounding the cylinders. Heat is extracted from the coolant by "radiators" through which ambient air is blown by means of a fan; for larger engines, the coolant may be brought

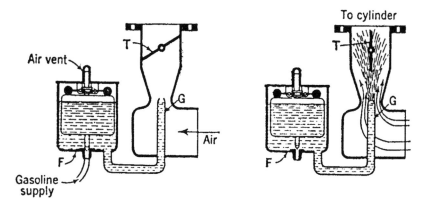

Figure 5-3. Plain tube carburetor; left, engine idle; right, engine in operation with wide-open throttle.

to outside cooling towers or heat exchangers in which water is circulated. More will be described later when discussing steam turbines.

Power Output and Speed

Engines are rated in terms of horsepower, usually at the maximum value of power and speed. (A horsepower is the power required to lift 550 pounds one foot in one second.)

The horsepower rating of an internal combustion engine required for driving a machine is usually greater than that of a steam engine or turbine driving the same machine. Internal combustion engines generally have no overload capacity, whereas a steam engine or turbine operate successfully at considerable overloads since they are usually rated on the horsepower at which the optimum steam economy is obtained. In either case, efficiencies (that is, the percentage of the heat value contained in the fuel is utilized) are low, being in the nature of 20 percent for gasoline and light fuel engines and 35 percent for heavy oil engines. While steam engines and turbines, in themselves, have a greater efficiency, in the nature of 75 percent, these values, when coupled with the boiler efficiencies that may vary from 30 to 70 percent, reduce the overall values that theoretically could range from about 20 to 50 percent, but in practice range from about 20 to 35 percent.

Speed ratings for internal combustion engines are taken at the maximum horsepower for which they are designed.

RECIPROCATING STEAM ENGINES

As a prime mover, the reciprocating steam engine has certain characteristics:

1. Ability to start under load.
2. Nearly constant and comparatively slow rotating speed.
3. Variable rotating speeds.
4. Direction of rotation may be reversed.

All of these characteristics can be designed in a single machine, but more often not all of them are incorporated in practice. Because of the relatively slow speeds and limited steam capacity, engines of this type are limited to about 2000 horsepower. For the production of large amounts of electrical energy, and for the higher speeds desirable in alternating current generation, steam turbines surpass the steam engine in economy of operation.

Simple Engine

The simple engine generally has only a single steam cylinder, although other cylinders may also be used. Steam is supplied to the cylinder where it is permitted to expand. The expansion of the steam converts the heat energy to mechanical energy by pushing on a piston through which, by means of a connecting rod and crankshaft, its motion is translated into a rotating mechanism capable of doing work.

The simplified type of reciprocating steam engine is known as the Slide-D-valve engine, using but one steam cylinder, and may be either horizontal or vertical. Figure 5-4 (a) and (b) represents this type of engine. The D-slide-valve alternately admits steam to and releases the steam from each end of the cylinder. The valve has a reciprocating motion. The piston is moved by the pressure of the steam admitted to the cylinder. At some point of the piston stroke the valve completely closes the supply port and the flow of steam is cut off. After cut off has taken place the pressure within the cylinder decreases owing to the expansion of the steam Energy stored in the heavy flywheel during the period of steam admission is used to maintain the stroke. Figure 5-5.

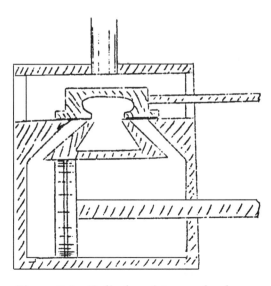

Figure 5-4a. Cylinder-piston and valve.

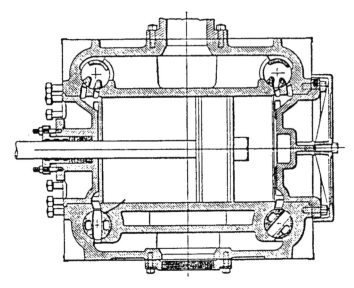

Figure 5-4b. Non-releasing engine cylinder.

Near the end of the piston stroke, the valve moves to uncover the exhaust port. The exhaust of expanded steam occurs while live steam is being admitted to the opposite end of the cylinder and continues until near the end of the return stroke of the piston. When the valve has moved to close completely both ports of the cylinder, some expanded steam remains ahead of the moving piston and is compressed with a rise in pressure. This compression acts as a cushion for the returning piston and continues until the valve has moved to admit live steam to that end of the cylinder. One such complete cycle takes place in each end of the cylinder per revolution.

The reciprocating motion imparted to the piston by the alternate admission, expansion, and release of steam in each end of the cylinder is transmitted by the piston connecting rod to produce rotary motion at the crankshaft.

The capacity of this simple engine may be increased by having steam act on both sides of the piston, in a double action operation. In this case, two D valves are employed, one acting for each side of the piston.

Other types of valves are also in use, but the general operation of such engines is unchanged. Figure 5-6.

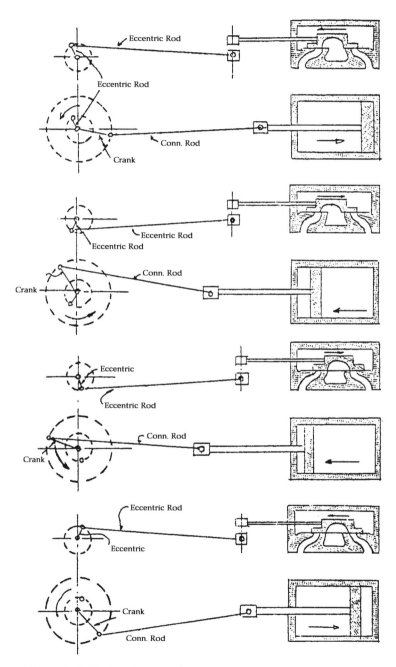

Figure 5-5. Cycle of operation—reciprocating steam engine.

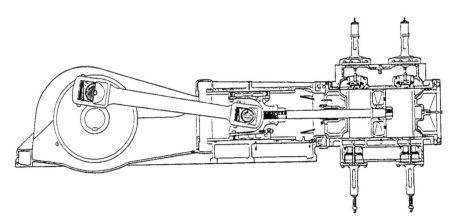

Figure 5-6. Longitudinal section of a poppet-valve steam engine.

Multiple Expansion

In a single expansion engine, such as the one described above, the steam expands in only one cylinder and is then exhausted.

In a multiple expansion engine, the steam is expanded a number of times or into a number of separate cylinders:

Compound — Two cylinders for expansion.

Triple — Three cylinders for expansion

Quadruple — Four cylinders for expansion.

In multiple expansion engines, steam is exhausted from the first cylinder and permitted to enter the second cylinder, where it is further expanded, and so on for the number of cylinders used. Higher thermal efficiencies are achieved, smaller flywheels can be used, and less condensation results; increased first cost and maintenance charges, however, are increased.

Cooling

Because of the relatively low temperatures involved in the operation of these engines, cooling presents only minor problems. Often the engines are cooled by the air of the ambient; larger units may have jackets similar to the internal combustion engines, with similar treatment of the coolant.

In smaller units, the spent steam is exhausted to the atmosphere. In larger units, the spent steam is passed through a condensing device and the condensate pumped back to the boiler to help replenish the water lost in the steam cycle. Condensers will be described later in connection with steam turbines,

STEAM TURBINES

Since the steam turbine can be built in units very much larger in capacity than is practical with reciprocating engines, it is almost universally in use in electric generating plants of even moderate sizes.

Simple Turbine

A simple turbine resembles a water wheel turned by a stream of flowing water; the falling water impinging on the paddles or vanes of the wheel imparts its energy to the wheel causing it to rotate and do work. The greater the flow of water, the greater the speed of rotation of the water wheel; if the speed of the water wheel is held constant, the greater flow accomplishes greater amounts of work. Figure 5-7a.

In a turbine, the steam under pressure is ejected from a jet and impinges on the blades of a wheel causing rotation of the shaft to which they are attached. The greater the flow of steam, the greater the forces exerted on the blades with a resulting greater speed or greater work performed. The steam expands, losing some of its energy in pushing on the blades; the unexpended portion of the steam escapes into the surrounding atmosphere. Figure 5-7 (b).

Stages

If the blades are curved, as shown in Figure 5-8 (a) and (b), the flow of steam, after impinging on one series of blades, instead of escaping into the surrounding area, can be redirected to impinge on a second set of blades, making use of more of the energy contained in the steam jet as it continues to expand. The redirected flow of steam expands in volume and loses some of its energy, and is made to impinge on an adjacent set of blades which is stationary. Here the steam is again redirected, this time impinging upon another movable set of blades, with more of the energy utilized. This process is continued until the resultant pressure and velocity of the steam becomes too low to be

Figure 5-7a. Hero's   **Figure 5-7b. Simple velocity turbine.**
reaction turbine.

practical. Each set of moving and stationary blades is called a *stage*; i.e., 15-stage, 30-stage, etc. turbine.

The number of stages that a turbine contains must be such that the energy-converting (to mechanical energy) capacity of the rotor is as nearly equal as practical to the available energy of the steam. The energy-converting capacity is determined by the diameter of the rotor, its length, revolutions per minute, the number and design of the stages.

Impulse Turbine

The design of the stages can be such that the expansion of the steam which passes through them occurs almost entirely in its stationary nozzles and its fixed or stationary blades; practically no expansion takes place in the moving blades. Such a design is called an impulse turbine; it is also called a velocity or equal pressure turbine. Figure 5-9.

Reaction Turbine

Another design, called a *reaction* turbine is one so designed that about half of the expansion of the steam which passes through it takes

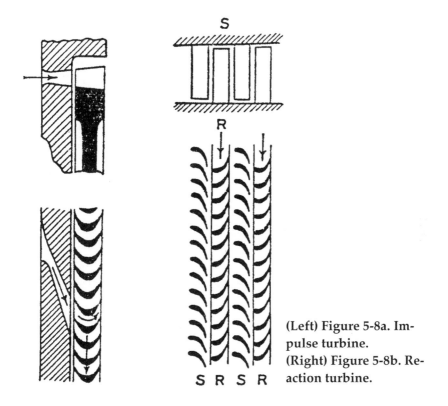

(Left) Figure 5-8a. Impulse turbine.
(Right) Figure 5-8b. Reaction turbine.

place in the moving blades and the other half in the guide or stationary blades. It is also called an unequal pressure turbine. Figure 5-10.

Impulse- Reaction Turbine

An *impulse* and *reaction* turbine is one which has some of its blades designed for impulse and some for reaction. The impulse blading is used in the first stages of the turbine, while the reaction blading is used in the rest of the turbine. Figure 5-11.

Blading

So that the energy converted to mechanical energy by each stage of the turbine be as nearly as practical the same, the diameter (and hence the length of the blades) of the several stages becomes progressively larger as the steam loses pressure in its progress through the stages. This tends to eliminate the stresses in the shaft of the rotor that would occur if the energy converted in each stage was significantly different,

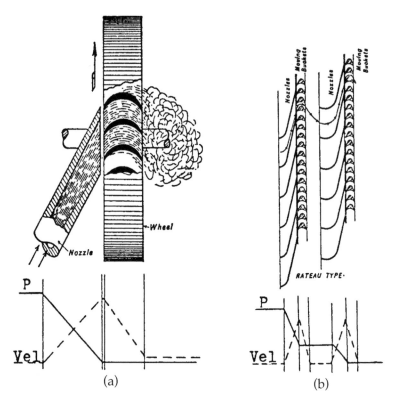

Figure 5-9. Impulse Turbines. a) Single stage—single entry. b) Pressure staged.

and results in a smooth rotation.

The blades of the turbine, being subject to superheated steam, corrode very little, but may expand and contract because of temperature variations. The moving blades may be made of nickel-steel or stainless (chromium) steel, while the stationary blades may be made of brass; all of these metals are high strength, heat resistant, and have relatively low coefficients of expansion.

The casing covering the rotor is shaped to accommodate the different size wheels, and also is designed to hold the stationary blades. The spent steam leaves the turbine casing at its widest point. Special precautions are required to prevent unequal expansion of the casing; ribs are provided so arranged to counteract any tendency to distortion due to the unequal temperatures at the inlet and exhaust ends.

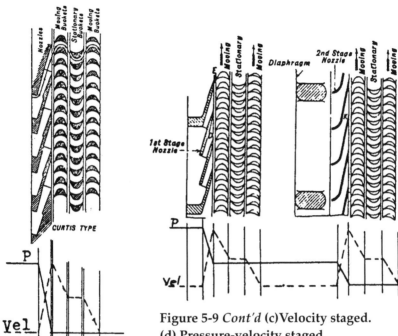

Figure 5-9 *Cont'd* (c)Velocity staged.
(d) Pressure-velocity staged.

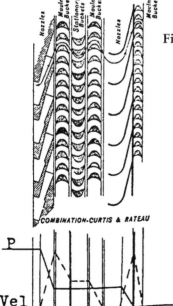

Figure 5-9e. Composite staged.

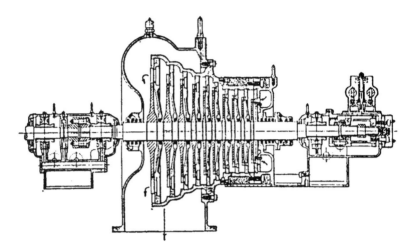

Figure 5-9f. Longitudinal section, impulse turbine with pressure staging; Rateau type.

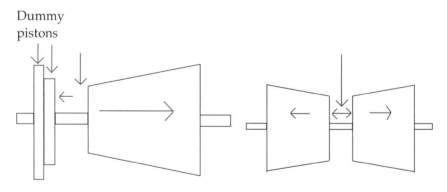

Dummy
pistons

Figure 5-10a. Single flow.

Figure 5-10b. Double flow.

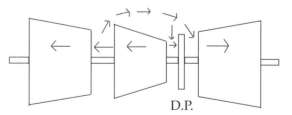

D.P.

Figure 5-10c. Single-double flow.

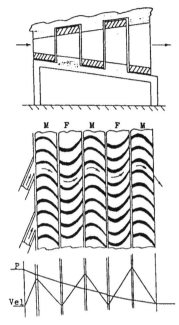

Figure 5-10d. Reaction turbine blading.

Special shaft packing glands are provided at the admission and exhaust ends of the turbine and are normally under vacuum with a continuous supply of a small amount of live steam to prevent air from entering the turbine. The main bearings are supplied with oil under pressure and a thrust bearing at the high pressure end takes up any uncompensated axial thrust.

The steel shaft of the turbine carries a heavy mass of metal and is also subject to expansion and contraction because of the variations in heat to which it is exposed. If the steam is suddenly discontinued, the heat contained in the large masses of metal will take time to dissipate, allowing the various parts to cool slowly. The heavy mass supported by the shaft will, however, cause it to bend, introducing a warp in the shaft which could damage the blades and other parts of the turbine. Provision is made, therefore, by means of a turning gear, to keep the rotor rotating until it is sufficiently cooled, a process that may take several hours. The shaft, however, will tend to bend even when stopped because of the huge mass it is supporting and similar precautions are taken when starting up, using the turning gear to rotate the rotor at gradually greater speeds until it is ready to pick up its load, a process that may also take several hours.

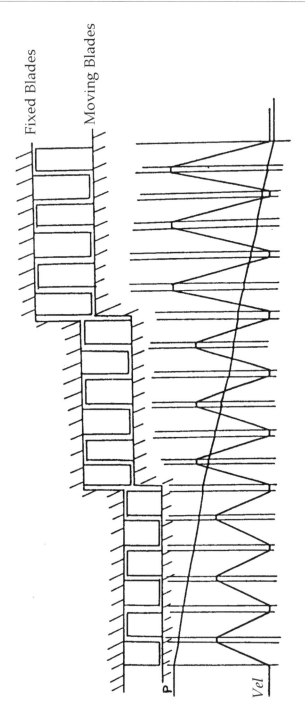

Figure 5-10e. Reaction turbine pressures and velocities.

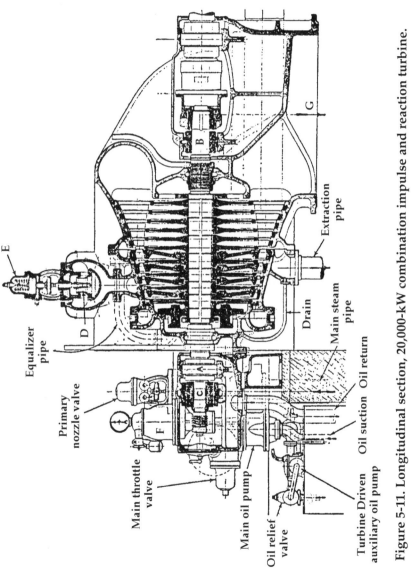

Figure 5-11. Longitudinal section, 20,000-kW combination impulse and reaction turbine.

Steam Flow in Turbines

For large turbines, the steam jets are generally designed so the steam flow is in a direction approximately parallel to the rotor shaft; these are known as *axial-flow* turbines. Small turbines have their steam flow approximately tangent to the rim of the rotor and are called *tangential-flow* turbines. The steam flow in smaller units may be radially inward toward or outward from the shaft; these are known as *radial-flow* turbines.

A turbine in which nearly all the steam that drives the turbine flows through the blades in the same general direction parallel to the rotor axis is known as a *single-flow* turbine. A turbine in which the main steam current is divided and the parts flow parallel to the rotor axis in opposite directions is known as a *double-flow* turbine. The latter is used to drive a large generator where the size of a single-flow turbine is so large as to become impractical or uneconomical.

Where generation requirements are large, resort is sometimes had to a *topping* turbine. This turbine operates at high pressure (1000 to 1200 pounds per square inch) and the entire volume of steam passes through it driving a generator. The steam exhausts at low pressure (500 to 600 pounds per square inch). This is then passed through one (or more) single-flow or double-flow turbines, driving a second (or third) generator of smaller capacity than the one driven by the high pressure turbine.

Bleeding

In order to obtain greater overall efficiency of the power plant, provision is made on the larger turbines for the extraction of steam at various stages for use in feedwater evaporators, heaters, and deaerators. This method, for the more complete utilization of the heat of the steam, is known as the regenerative cycle, and results in substantial reduction in the heat consumption of the plant as a whole. For example, a turbine with a total of twenty stages may receive steam at about 400 pounds per square inch: steam from the tenth stage at about 80 pounds per square inch goes to feedwater heaters and evaporators; steam from the fifteenth stage at about 30 pounds per square inch goes to feedwater heaters and deaerators; and from the eighteenth stage at about 6 pounds per square inch to a feedwater heater. The amounts of steam extracted at each of the "bleed points" may vary in accordance with the need. Figures 5-12 (a), (b).

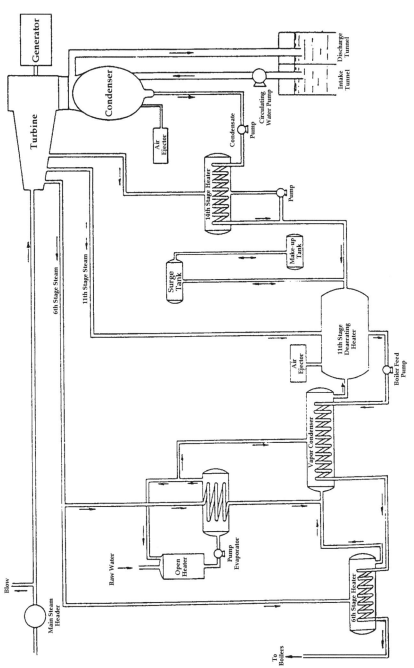

Figure 5-12a. Schematic diagram of three stage turbine.

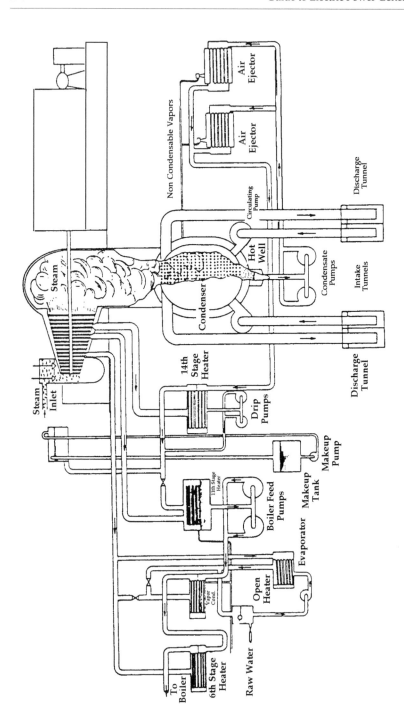

Figure 5-12b. Pictorial diagram of three stage turbine.

Steam Turbines vs Reciprocating Steam Engines

The reciprocating steam engine utilizes economically a greater amount of the available energy of the steam at high pressures than at low pressures. The turbine, on the other hand, has the advantage at low pressures and in cases where large expansion of the steam are uses. To obtain the same steam expansion in a reciprocating engine, the cylinders would be very large with greater cylinder losses. The greater expansion of steam in turbines permits the recovery of a large amount of power from what otherwise (in the reciprocating steam engine) would be waste steam.

The most important energy losses that occur in a turbine include: steam leakage, wheel and blade rotation, left-over velocity of the steam leaving the blades, heat content of the exhaust steam, radiation, and friction.

The steam consumption of the reciprocating engine and the turbine are approximately the same, but the steam turbine operates on the complete expansion cycle, the reciprocating engine operates on a cycle with incomplete expansion. The turbine, therefore, converts into work a greater amount of heat per pound of steam used.

Although steam pressures and temperatures, and turbine designs continue to improve the overall efficiencies of power plants, nevertheless the thermal efficiency of steam turbines hovers around 35 percent. The steam turbine, however, is well adapted to the driving of electric generators of the alternating current type as the desirable speeds of the two machines are the same, and is almost universally used in steam plants larger than about 1000 kW capacity.

Condensers

As their name implies, condensers are employed to help condense spent steam into water. In this process they act as heat exchangers, extracting heat from the steam in lowering its temperature and further extracting its heat of vaporization in turning the steam into water. One other result of this action is to cause a vacuum at the exhaust end that helps the steam to pass through the turbine.

When engines exhaust steam to the atmosphere, the exhaust pressure is atmospheric or about 14.7 pounds per square inch. When they exhaust in a condenser, the action is such that the back pressure is reduced to some value below atmospheric pressure. The difference between atmospheric pressure and the *absolute* pressure in the condenser is known as a vacuum. (Absolute pressure is the pressure above zero pressure; it is the gauge pressure plus the atmospheric pressure, both expressed in the same units.) The effect of this vacuum is to increase the effective pressure difference between the pressure of the incoming steam and that of the exhaust steam; this results in a considerable saving in the steam requirements to achieve the same results if no condenser was employed.

Types of Condensers
Condensers may be divided into two different classes:

1. Jet condensers, in which condensation of the exhaust steam is affected by direct contact of the vapor and the cooling water.

2. Surface condensers, in which the exhaust steam and the cooling water do not come in direct contact with each other; the heat is extracted from the steam by heat transfer through metal walls or tubes.

The jet condenser cools the incoming exhaust steam by causing it to flow under a fine water spray. The mixture of condensate and cooling water also contains air which is carried along with the mixture and removed from the bottom of the condensing chamber. In one type, referred to as a low-level jet condenser, this residue is removed by means of a pump. See Figure 5-13 (a). In another type, known as a barometric condenser, the air is removed by a vacuum pump above the shower head; the condensed steam and water flow down into a "tail" pipe and the water vapor collected in the air chamber flows into a drain pipe, both emptying into a so-called hot-well. The tail pipe is a minimum of 34 feet in length. As atmospheric pressure can support a column of water of about this same height, as soon as the water reaches this height it flows into the hot-well, performing the same function as a pump. Figure 5-13 (b). Water from the hot-well is returned to the feedwater system.

The surface condenser does not permit the cooling medium and the exhaust steam to come into contact with each other. In the usual surface condenser, the cooling water circulates through tubes, and the steam is brought into contact with the outer surface of the tubes. The water in the cooling tubes may come from bodies of water, such as riv-

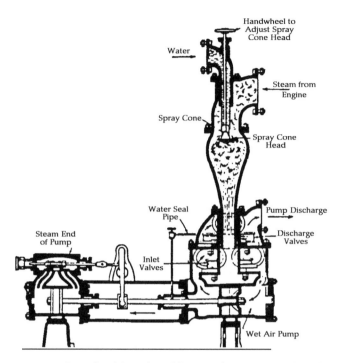

Figure 5-13a. Standard low level jet condenser—sectional view.

ers, lakes or oceans whose purity and salinity, although not affecting the steam, is usually taken into account in the design and operation of condensers. Figures 5-14 (a) and (b).

Where such bodies are unavailable or too small, the cooling water may be circulated through sprays in ponds, or flowing along the surfaces of cooling towers where the surrounding air cools the water before it is circulated back to the condenser. The condensate can be used for boiler feedwater and its temperature should be lowered to as high a value as practical. In many installations, this type of condenser is situated just below the turbine. In another type, the steam is circulated in pipes immersed in the cooling water which is cooled in the same manner as described above; the tanks in which the water is contained can be entirely closed, or open to the atmosphere.

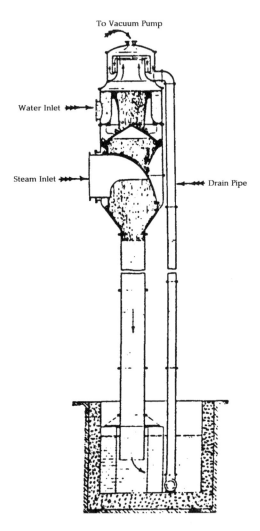

Figure 5-13b. Barometric condenser.

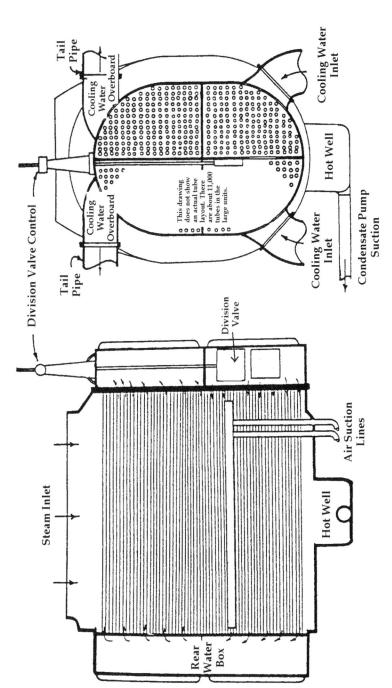

Figure 5-14a. Surface condenser sectional views.

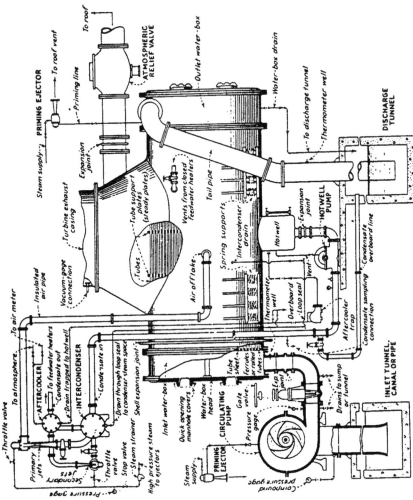

Figure 5-14b. Surface condenser connections.

Comparison of Condensers

Advantages	Disadvantages

Jet Condensers
1. First cost low.
2. Less floor space.
3. Low maintenance.
4. Uses clean water.

1. All the condensate is lost.
2. Pumping requirements high.
3. Increased cost due to using raw Water for feedwater.

Surface Condensers
1. Condensate is separate.
2. High vacuum attainable.
3. Low pumping requirements.

1. First cost high.
2. Maintenance is high.
3. Large space is required.

Governors

The purpose of engine governors is to maintain approximately constant speed of rotation and to obtain optimum thermal efficiency with varying loads. The governing of prime movers or engines, whether of the hydro, internal combustion, reciprocating steam, or turbine types does not differ materially. The control is accomplished automatically by mechanical and electrical types of governors.

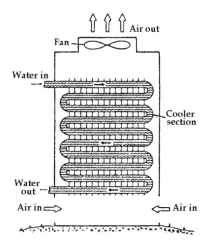

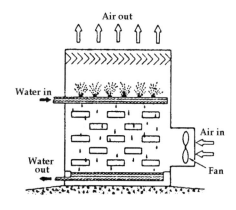

Figure 5-16a. A dry cooling tower system in which the flow of air is provided by mechanical means.

Figure 5-16b. A wet cooling tower system in which the flow of air is provided by mechanical means.

Mechanical Types
Centrifugal Fly-Ball Type

The simplest type of governor is known as the "fly-ball" type. The governor action depends upon centrifugal force to raise and lower the rotating balls as the speed increases or decreases; see Figure 5-16 (a). An increase in speed will increase the centrifugal force thereby raising the collar 'A' to some other position 'B'. Directly connected to this rising collar is a linkage through which the throttle valve is moved to a smaller opening thereby cutting down the fuel or steam supply to the engine or turbine. The reverse operation takes place when the speed is decreased: the reduction in centrifugal force allows the collar to drop, opening the throttle valve. In the case of internal combustion engines, the linkages may also control the mixture of fuel and air for optimum efficiency. Other types are shown in Figures 5-16b, c, and d.

Inertia-Centrifugal Eccentric

Sometimes known as a "shaft governor," it is commonly located in the flywheel of the engine and governs the speed by changing the position of the eccentric. Figure 5-17 shows the essential parts of a typical eccentric governor.

A weight arm wW is pivoted at P and is constrained by a spring;

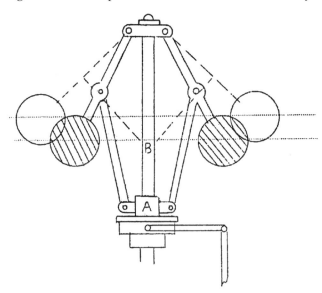

Figure 5-16a. Centrifugal Fly-ball Type Governor.

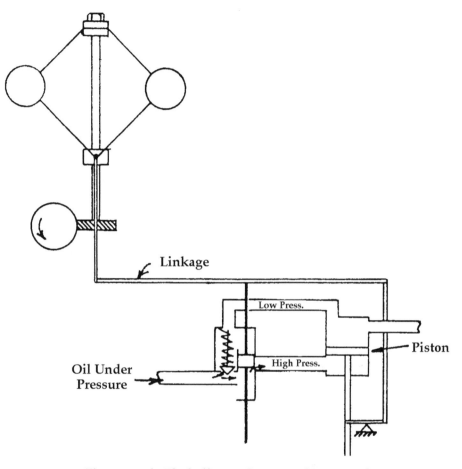

Figure 5-16b. Fly-ball type Governor (*Continued*)

the arm carries the eccentric pin E and the two weights w and W. As the flywheel rotates, the arm tends to move in the opposite direction. The weight W will swing out toward the rim of the flywheel and the eccentric pin E will swing toward e. The movement of the arm can be regulated by the tension of the spring. An increase in speed will cause the weight W to move outward by its inertia and centrifugal force, and the eccentric pin E will move toward e, and in so doing changes the angle of advance and eccentricity of the valve gear. A decrease in the eccentricity will give a smaller travel to the valve and the increase in the angle of advance will shorten the cut-off, both actions are such that the power is cut down to counteract the increased speed thereby slowing down the engine.

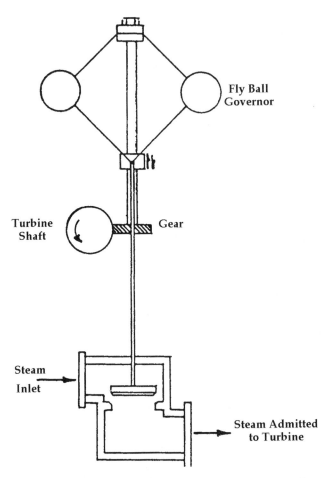

Figure 5-16c. Fly-ball type Governor (*Continued*)

Electrical Type
Contact Making Voltmeter

The simplest type of electrical speed control is the "contact making voltmeter." A small direct-current generator is mounted on the shaft of the engine, the voltage generated will depend on the speed. At the rated speed, the voltmeter indicator will float between two contacts. When the speed of the engine slows down, the voltage generated will be lower and the voltmeter indicator will make contact with one of the contacts which, through auxiliary relays, will cause the motor operating the throttle to open wider restoring the speed of the engine. A similar action takes

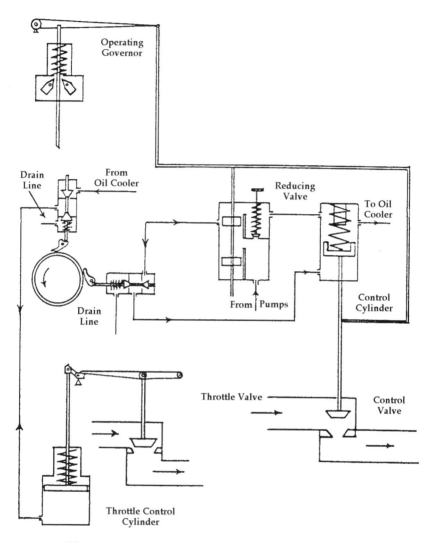

Figure 5-16d. Typical governor control mechanism.

place when the speed is too high; the other contact is actuated which will cause the motor operating the throttle to close restoring the speed of the engine. The spread in voltage values necessary to make contacts between the 'high' and 'low' limits, known as the 'band,' cannot be too small as these contacts will "chatter" and, since, because of the mass of the generator will take some time to adjust its speed, the contacts will

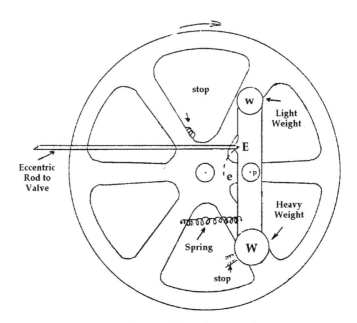

Figure 5-17. Inertia-centrifugal eccentric governor.

cause undesirable changes in the speed of the engine. The "band" must be wide enough to prevent this chatter, but may result in undesirable variations in speed. Figure 5-18.

Self-Synchronization (Selsyn)

A more complex type pits the voltage generated by the shaft generator against a voltage generated by a standard unit. When speeds are "synchronous," no current will flow in the circuit regulating the throttle motor. As the speed of the engine begins to vary (faster or slower), the balance is upset and a current will flow causing the throttle motor to operate to restore the synchronism. Unlike the contact making voltmeter, the synchronizing action is continuous and results in very fine variations and smooth rotation of the engine. Figure 5-19.

Another variation pits the frequency generated by the engine shaft is pitted against a quartz (or other) crystal which will allow current to flow when the frequency of the current varies. Through electronic circuitry, relays operate to keep the throttle setting at the prescribed value. The action here is even finer and results in a practically constant speed of the engine.

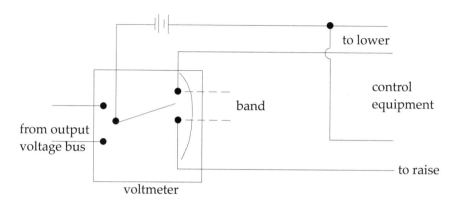

Figure 5-18 Contact-making voltmeter control.

Because of the sensitivity of these devices, they are associated with control circuitry that will shut down the machine should these devices fail for any reason.

Heat Balance Diagram

A Heat Balance Diagram for a three stage turbine driven power plant is shown in Figure 5-20 (a). Pounds of steam, pressures and temperatures at each stage are indicated.

A Heat Balance Diagram for a turbine driven generator, a by-product of industrial use steam, is shown in Figure 5-20 (b).

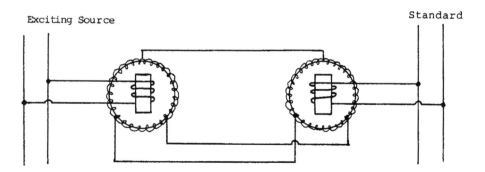

Figure 5-19 Diagram of Selsyn Set with interconnected stator windings and rotors connected to exciting source and standard.

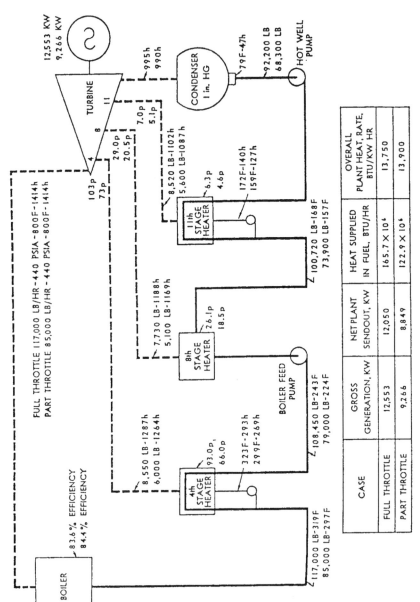

Figure 5-20a. Heat balance diagram for 10,000 kW steam power plant.

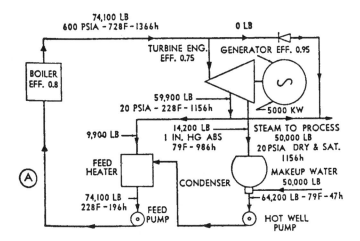

Figure 5-20b. By-product generation of electric power and process steam net electric sendout 4650 kW process steam 50,000 lb/hr.

THE GAS TURBINE

The gas turbine produces mechanical work using gas as the working medium. It is very similar to the steam turbine; the expanding gas provided by the ignition of fuel, through nozzles, impinge on the blades that are attached to a shaft which is made to rotate, producing work. Like the steam turbine, the gas may be introduced in stages depending on the temperature to improve operating efficiency and the metals involved subjected to higher temperatures. (Refer to figure 5-21)

Comparative features of the gas turbine as a prime mover are:

1. It is simpler than the steam turbine and internal combustion engine.

2. It requires an independent source for starting and bringing up to speed.

3. Like the steam turbine, it is not readily able to reverse direction of rotation.

4. Attained speeds are higher and with less vibration than engines of similar size.

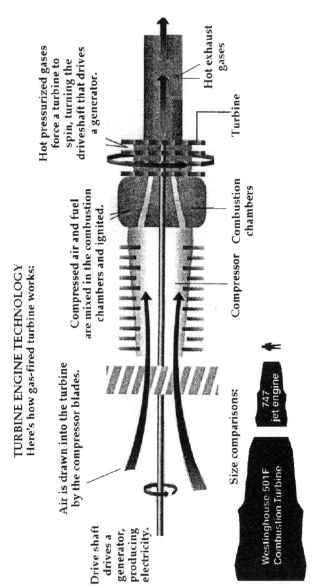

Figure 5-21a. Turbine Engine.

Figure 5-21b. Microturbine.

5. Because of higher temperatures involved, maintenance requirements may be greater to achieve comparable service work.

6. Compression of air supply detracts from output and overall efficiency.

Fuels used are generally oil or natural gas. Since products of combustion flow though the turbine, solid coal (producing ash) cannot be used, but gasifies (as in syngas) makes it competitive with other fuels.

Chapter 6

Generators

G enerators of electricity depend on magnetism for their operation, a phenomenon with which everyone may be familiar. Before exploring the design and operation of generators further, it is desirable, although not essential, to have a basic understanding of the phenomenon of magnetism and its association with electricity.

ELECTRO-MAGNETISM

This form of magnetism is the same as found in a natural magnet and in so-called permanent magnets similar to the toy and novelty variety familiar to most everyone. Electro-magnetism, however, is created artificially and ceases to exist when the creating force is removed.

ELECTROMAGNETIC FIELDS

A current of electricity flowing through a wire produces not only heat, but also a magnetic field about the wire. This is proved by placing a compass needle in the vicinity of the current carrying wire.

If the direction of the electric current is assumed to be from positive to negative, then it will be observed that the magnetic needle placed adjacent to the conductor will always point with its "North" pole in a certain definite direction. The needle is forced into this position by what is known as magnetic lines of force, sometimes also referred to as magnetic flux.

Right Hand Thumb Rule

This observation leads to a general rule known as the right hand thumb rule. Imagine that the wire is grasped in the right hand with the thumb out stretched in the direction of the electric current. Then the direction in which the fingers curl around the wire is the direction of the magnetic lines of force.

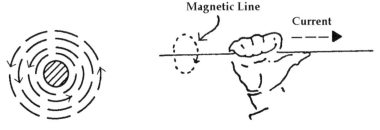

Figure 6-1. Magnetic field about a conductor.   **Figure 6-2. Right hand thumb rule.**

Field About a Coil

The magnetic field around a single wire carrying a current is rather weak. By winding the wire into a ring, the magnetic lines are concentrated in the small space inside the coil and the magnetic effect is much increased. The grouping of the lines of force is known as a magnetic field.

A coil of wire is nothing but a succession of these rings stacked one after the other. Each adds its quota to the magnetic field. Most of the

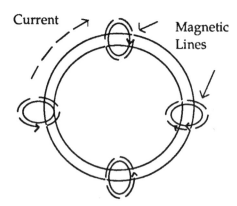

Figure 6-3. Magnetic field about a wire loop.

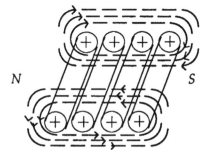

Figure 6-4. Magnetic field about a coil.

magnetic lines of force pass straight through the coil. Each line makes a complete circuit, returning by a path outside the coil. A coil carrying a current is in fact a magnet. Where the lines come out is the "North" pole; where they enter if the "South" pole.

Right Hand Rule for Coil Polarity

The polarity of the coil may be determined by the "right hand rule." If the coil is grasped in the right hand with the fingers pointing in the direction the current is flowing around the coil, the outstretched thumb points toward the "North" pole of the coil.

The strength of the magnetic field inside a coil depends on the strength of the current flowing and the number of turns. It is therefore expressed in "ampere-turns," that is, amperes multiplied by the number of turns. Thus a single turn carrying a very large current may produce the same effect as a great many turns carrying a small current.

A coil with an air-core, however, produces a comparatively weak field. The strength is enormously increased by putting in a core of soft iron. This is generally referred to as an electromagnet.

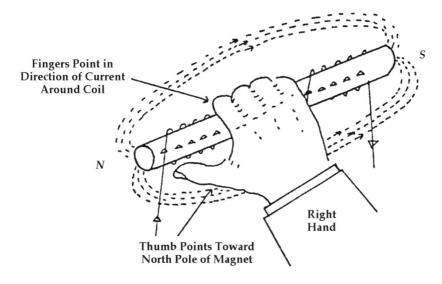

Figure 6-5. Application of right-hand rule to determine polarity of coil.

ELECTROMAGNETIC INDUCTION

One relation between electricity and magnetism has already been demonstrated, that of producing magnetism with the aid of electric current. There is another most important relation and that is the production of electricity with the aid of magnetism.

When a conductor is moved through a magnetic field an electrical pressure—or voltage—is produced in the conductor. It requires work to push the conductor through the magnetic field. In this case, the magnetic field acts as a force which resists the movement of the conductor. The energy which is used in pushing the conductor through the magnetic field is equal to less losses the electrical energy generated—or "induced"—in the conductor. This simple fundamental provides a means of converting mechanical work directly into electricity.

GENERATOR ACTION

This phenomenon is illustrated in Figure 6-6a, which shows a magnetic field established in the air gap between the North and South poles of a magnet. Through this field extends a conductor. If the conductor is moved downward between the two poles of the magnet, an electrical pressure of voltage—is set up in it which attempts to circulate the electrons, that is; an electric current tends to flow (in the direction indicated but is prevented because the circuit is not complete. If the circuit is completed into a loop as shown below by the broken line, then a stream of electrons would circulate around the loop—or an electric

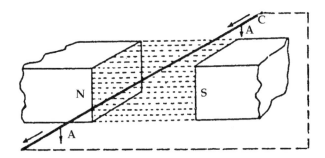

Figure 6-6a. Conductor moving in direction A-A through magnetic field.

current would flow in the loop. The current would flow only when the wire moved through the magnetic field. The movement of the wire would be resisted by the magnetic field just as though the two poles of the magnet were connected by many rubber bands.

Factors Affecting Induced Voltage

If the simple magnet is substituted by a more powerful electromagnet, it is found that a greater electrical pressure—or voltage—is now induced in the conductor cutting the magnetic lines of force. That is, the greater the number of lines of force for the stronger the magnetic field), the greater will be the voltage induced. However, more mechanical work is now required to move the conductors through the stronger magnetic field.

The greater the length of the conductor, the more voltage is produced because more lines of force are cut. If the conductor cuts the magnetic lines of force at an angle, it will cut only the same number of lines of force as when it is cutting them at right angles. In considering conductor length, therefore, only the effective length, that is length that cuts the lines of force at right angles, should be considered.

Similarly, if the speed at which the conductor is moved through the magnetic field is increased, it is found that the voltage induced in the conductor is also increased. Here, too, more mechanical work is required to move the conductor at greater speed through the magnetic field.

The magnitude of the electrical pressure induced in a conductor while it is moving through a magnetic field therefore, is determined by the rate of cutting of the lines of force of the magnetic field. The rate of cutting of the lines of force depends on three factors:

1. The length of the conductor which is cutting through the magnetic field.

2. The speed at which it is moving.

3. The strength or density of the magnetic field—that is, the number of lines of force per square inch.

The cutting of magnetic lines of force by a conductor moving through a magnetic field may be compared to the cutting of blades of grass on a lawn. Three factors that determine the rate of cutting of the blades of grass are:

1. The length of the cutting bar of the mower at right angles to the swath cut.
2. The speed at which the cutting bar moves through the lawn.

3. The density of the lawn—that is, the number of blades of grass per square foot of lawn.

A change in any of these three factors makes a corresponding change in the rate of the cutting of the blades of grass as well as in the amount of work required to cut the grass. If a two foot cutting bar is used, the rate of cutting as well as the work required will be one-third more than if a one and a half foot cutting bar were used. Here it will be noted that less grass will be cut if the mower is pushed so that it is not at right angles to the swath cut. If the speed at which the mower is moved through the lawn is doubled, the rate of cutting is doubled as is also the work necessary to push the mower twice as fast. If the mower is moved through a lawn in which there are many blades in each square foot of lawn, the rate of cutting and the work required are correspondingly greater than in a less dense lawn in which there are fewer blades of grass in each square foot area.

Relative Motion of Conductor and field

An electrical pressure—or voltage—can be induced in a conductor by moving it through a magnetic field as described above. A voltage can also be induced electromagnetically by moving a magnetic field across the conductor. It makes no difference whether the conductor is moved across the magnetic field or the magnetic field is moved across the conductor. A stationary conductor which has a magnetic field sweeping across it is cutting the magnetic field just the same as though the conductor were moving across the magnetic field.

Referring to figure 6-6(b), it is observed that the conductor moving downward through the magnetic lines of force whose direction is assumed to be from the North pole to the South pole of the magnet has a voltage induced in it such that the circulation of electrons—or electric current—is in the direction indicated. If the conductor is moved upward through the same magnetic lines of force, the voltage induced in it will be in the opposite direction to that indicated above. Thus, it is seen that

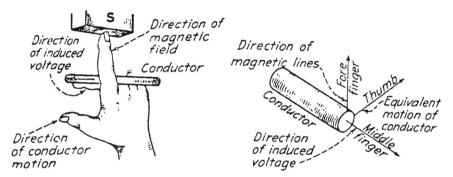

Figure 6-6b. Right hand rule for generation action.

the direction of the induced voltage depends upon the direction of the magnetic field and the direction of motion of the conductor.

Right Hand Rule

It is convenient to have a rule for remembering the relations in direction between the magnetic field, the motion of the conductor and the voltage induced in it. This rule is usually called the Right Hand Rule. If the right hand is held with the thumb, forefinger and middle finger all at right angles to each other as shown in figure 6-6, the thumb indicates the direction of motion of the conductor through the magnetic field, the forefinger indicates the direction of the magnetic lines of force (issuing from the North pole) and the middle finger indicates the direction in which a voltage is being generated in the conductor. The middle finger also indicates the direction in which the current will flow in the conductor. See Figure 6-6(b)

MOTOR ACTION

Mechanical work can be converted into electrical energy by pushing a conductor through a magnetic field. Conversely. If a conductor carrying an electric current is placed in a magnetic field, there will be a force produced tending to move the conductor. This moving conductor can be harnessed to do some mechanical work.

The same general relations that exist between the magnetic field motion of the conductor and the electrical pressure or current in the conductor also apply to this latter phenomenon. This is known as motor action to distinguish it from the first phenomenon known as generator action. A rule for motor action uses the left hand, and the quantities represented by the fingers remain the same as for the generator Right Hand Rule, described above. See Figure 6-6(c)

Electrical Quantities

Electrical *pressure* is expressed in *volts* (or kilovolts equal to 1000 volts), analogous to water pressure expressed in pounds per square inch.

Electrical *current* is expressed in *amperes*, analogous to water current expressed in cubic feet (or gallons) per minute, or hour.

Electrical *resistance* is expressed in *ohms*, analogous to friction encountered by a flow of water in a pipe, for which no separate unit is denoted but is expressed in loss of water pressure or reduction in current flow. In an electric circuit, resistance has a similar loss of electrical pressure in volts:

Pressure drop (volts) = Current (amperes) × Resistance (ohms)

Ohm's Law expresses the relationship that exists between these three electrical quantities, namely: the flow of current varies directly with the pressure and inversely as the résistance, or

$$\text{Current (amperes)} = \frac{\text{Pressure (volts)}}{\text{Resistance (ohms)}}$$

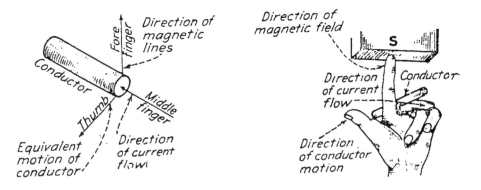

Figure 6-6c. Left hand rule for motor action.

PROPERTIES OF ELECTRIC CIRCUITS

Electric Circuits
In every electrical circuit there must be a complete path for the flow of the current. This path extends from one terminal of the source of supply (usually designated as the positive terminal), through one of the conducting wires, through the device or devices using the energy, through the other of the conducting wires to the second terminal of the source (usually designated as the "negative" terminal), and back through the source to the first (or positive) terminal.

There are two fundamental types of electrical circuits known as the series circuit and the multiple or parallel circuit. Other types are a combination of these. The following discussion applies both to devices producing electrical energy (generators) and to devices receiving electrical energy (appliances).

SERIES CIRCUITS

In a series circuit all the parts that make up the circuit are connected in succession, so that whatever current passes through one of the parts passes through all of the parts.

Simple Series Circuit
The circuit shown in figure 6-7 contains four resistances, R_1, R_2, R_3, and R_4, which are connected in series. The same current (I) flows through each of these resistances. Assume an electrical pressure (E) of 100 volts is applied across the terminals of the circuit, and that a current of 5 amperes flows through the circuit. By applying Ohm's Law, the resistance (R) of the circuit as a whole is found to be 20 ohms; that is:

$$R = \frac{E}{I} \text{ or } \frac{100 \text{ volts}}{5 \text{ amperes}} = 20 \text{ ohms}$$

This total resistance is the sum of the resistances R_1, R_2, R_3. The current of five amperes is the same throughout the circuit. The volts across R_1, R_2, R_3, R_4 are measured and found to be:

$$E_1 = 15 \text{ volts} \qquad E_2 = 25 \text{ volts}$$

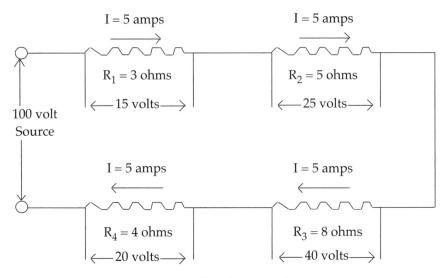

Figure 6-7. Simple series circuit.

$E_3 = 40$ volts $E_4 = 20$ volts

Now, Ohm's law applies to any part of the circuit as well as to the entire circuit. Applying Ohm's law to each resistance in the series circuit:

$$\text{Since } E_1 = 15 \text{ volts, } R_1 = \frac{E_1}{I} = \frac{15 \text{ volts}}{5 \text{ amperes}} = 3 \text{ ohms}$$

$$E_2 = 25 \text{ volts, } R_2 = \frac{E_2}{I} = \frac{25 \text{ volts}}{5 \text{ amperes}} = 5 \text{ ohms}$$

$$E_3 = 40 \text{ volts, } R_3 = \frac{E_3}{I} = \frac{40 \text{ volts}}{5 \text{ amperes}} = 8 \text{ ohms}$$

$$E_4 = 20 \text{ volts, } R_4 = \frac{E_4}{I} = \frac{20 \text{ volts}}{5 \text{ amperes}} = 4 \text{ ohms}$$

As a check, it is found that the sum of the four individual voltages equals the total voltage: $E_1 + E_2 + E_3 + E_4 = E$

15 volts + 25 volts + 40 volts + 20 volts = 100 volts

Also, the sum of the separate resistances is equal to the total resistance; that is:

$$R_1 + R_2 + R_3 + R_4 = R$$

3 ohms + 5 ohms + 8 ohms + 4 ohms = 20 ohms

When a current flows in a circuit, there is a continual drop in electrical pressure from one end of the circuit to the other. This drop is given by Ohm's law. It is E = IR and is usually known as the IR drop. In all leads and connecting wire, this drop is kept as small as possible because it represents a loss.

In every appliance through which a current is sent there is also an IR drop, usually much larger than in the connecting wires because the resistance of the appliance is higher. If several appliances are connected in series, the electrical pressure does not drop evenly around the circuit but by steps, each step representing an appliance. Since the current is the same in all of them, the IR drop in each appliance is proportional to its resistance.

The generator or source of electrical pressure provides a certain total voltage for the circuit. Each appliance spends a part of it, but all these parts must add up to the original pressure or voltage.

MULTIPLE OR PARALLEL CIRCUITS

A multiple or parallel circuit is one in which all the components receive the full line voltage, the current in each part of the circuit being dependent on the amount of opposition (resistance) of that part of the circuit to the flow of electricity.

The circuit shown in Figure 6-8 contains four resistances, R_1, R_2, R_3, and R_4, connected in multiple. Assume again that an electrical pressure (E) of 100 volts is applied across the terminals of the circuit. Here, each branch of the circuit receives the full voltage of 100 volts. Assume the resistances have the following values:

$$R_1 = 1 \text{ ohm}, R_2 = 4 \text{ ohms}, R_3 = 5 \text{ ohms}, R_4 = 20 \text{ ohms}$$

Again, applying Ohm's law to find the current flowing in each resistance:

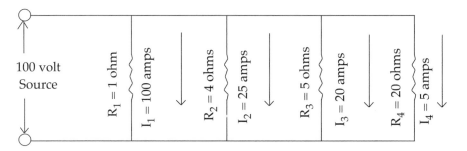

Figure 6-8. Resistances in multiple.

$$\text{Since } R_1 = 2 \text{ ohms, } I_1 = \frac{E}{R_1} = \frac{100 \text{ volts}}{1 \text{ ohm}} = 100 \text{ amperes}$$

$$R_2 = 4 \text{ ohms, } I_2 = \frac{E}{R_2} = \frac{100 \text{ volts}}{4 \text{ ohms}} = 25 \text{ amperes}$$

$$R_3 = 5 \text{ ohms, } I_3 = \frac{E}{R_4} = \frac{100 \text{ volts}}{5 \text{ ohms}} = 20 \text{ amperes}$$

$$R_4 = 20 \text{ ohms, } I = \frac{E}{R_4} = \frac{100 \text{ volts}}{20 \text{ ohms}} = 5 \text{ amperes}$$

The total current in a parallel circuit is equal to the sum of the separate currents:

$$I_1 + I_2 + I_3 + I_4 = I$$

100 amps + 25 amps + 20 amps + 5 amps = 150 amps

The resistance of the entire circuit may be found by applying Ohm's law:

$$R = \frac{E}{I} = \frac{100 \text{ volts}}{150 \text{ amps}} = \frac{2}{3} \text{ or } 0.667 \text{ ohm}$$

Another way of obtaining the resistance of the entire circuit is to add the reciprocals of each of the resistances and taking the reciprocal of the sum (the reciprocal of a number is equal to I divided by that number); that is

$$\frac{I}{R_1} + \frac{I}{R_2} + \frac{I}{R_3} + \frac{I}{R_4} = \frac{I}{R}$$

$$\frac{I}{1 \text{ ohm}} + \frac{I}{4 \text{ ohm}} + \frac{I}{5 \text{ ohm}} + \frac{I}{20 \text{ ohm}} = \frac{I}{R}$$

$$1.00 + 0.25 + 0.2 + 0.05 = 1.50 = \frac{I}{R}$$

Reciprocal of $\frac{I}{R}$ is R or $R = \frac{I}{1.50} = 0.667$ ohm

If resistances are connected in series, the ohms simply add up (as do the volts for generators connected in series), but, if they are connected in multiple it is observed that the resultant resistance is less than the smallest of the component resistances in the circuit. This is so because each additional resistance provides an additional path for the circuit, so that more current can flow. The conducting ability of the circuit is increased, the resistance is lowered.

It is also observed that connecting in series adds up the ohms (or volts); connecting in parallel adds up the amperes.

Series-Parallel Circuits

An example of resistance connected in series-parallel is given in Figure 6-9. To get the total resistance of this circuit, the resultant resistance of each of the two parallel groups is first determined, then the resistances of groups 1, 2, and 3 are added. The same process applies to any number and kind of group.

HEAT LOSS

By Ohm's law, E = IR. If IR is substituted for E in the expression for power,   W = EI

It becomes: W = I × IR

This may also be written $W = I^2R$

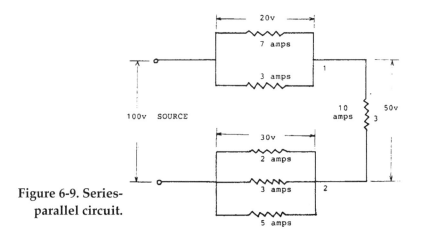

Figure 6-9. Series-parallel circuit.

where I^2, or I squared, signifies I multiplied by itself, that is, I × I.

Therefore, if the current flowing through a circuit is known, as well as its resistance, it is possible to determine the power necessary to overcome the effects of the resistance. Where this power does not produce useful work, that is, where the electrical energy is not converted to some mechanical work, it is converted into heat and dissipated into the surrounding atmosphere. This heat, owing to the electrical resistance encountered, may be likened to the heat developed by friction, and represents a loss.

An example of such a condition, is the heat loss in the wires which carry the electric current from the generator to, let us say, an electric motor. In determining R, however, the resistance of the wires of the motor must be included, for there are losses in these as well as in any other wires.

If the current in a wire is DOUBLED, the heat loss is therefore QUADRUPLED (not doubled), if the resistance remains the same.

Direct and Alternating Current

There is a resemblance between the flow of electricity in an electric circuit and the flow of water in a water system. The simple electric circuit illustrated consists of a generator, wire conductors, switch and a motor. In a water system illustrating a direct current circuit, the pump causes water to flow in one direction through pipes, valve and a water wheel. In a water system illustrating an alternating current circuit, a piston type pump causes water to flow back and forth (alternating directions) in the pipes, valve and a reciprocating type engine. In either case, a continuous performance (or rotation) results which can be kept up indefinitely. See Figure 6-10 (a), (b) and (c).

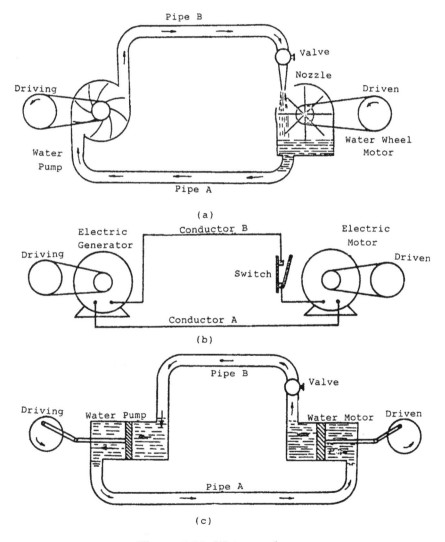

Figure 6-10. Water analogy.

ELECTRIC GENERATORS

Alternating Current Generators

Obviously, the intermittent production of an electrical pressure—or voltage—produced by moving a conductor up or down through a magnetic field as described earlier, is not suitable for practical purposes. What is desired is the production of an electrical pressure as a continuous performance.

A convenient and practical method of achieving this result is to mount the conductor between two insulated discs which may then be rotated by some outside mechanical machine. This is shown diagrammatically in Figure 6-11. The conductor is rotated in a magnetic field and the two ends are provided with slip rings and brushes enabling a connection to be made to an outside circuit.

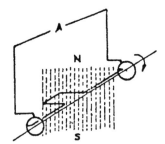

Figure 6-11. Single conductor rotating in magnetic field.

Single-Phase Generators

In Figure 6-12(a), (b), (c), and (d) is shown a series of diagrams in which the magnetic field is merely indicated. The conductor is assumed to be rotating with a uniform speed in the direction shown by the curved arrows. At "start" the voltage will be zero because the conductor is moving parallel to the magnetic field; no lines of force are cut when the conductor moves in a direction parallel to the magnetic field. Gradually, as the conductor revolves, it begins to cut more and more across the magnetic field. As it approaches the position marked "1/4 cycle," for each degree it moves, the number of lines of force that are cut becomes greater and greater. At the "1/4 cycle" point, it is cutting squarely across the magnetic field and the voltage in the conductor will be a maximum, but will reduce again as the conductor revolves to the "1/2 cycle" position, where it is zero. Now, as the conductor keeps revolving, it will be noted that the conductor will cut the magnetic lines of force in a direction

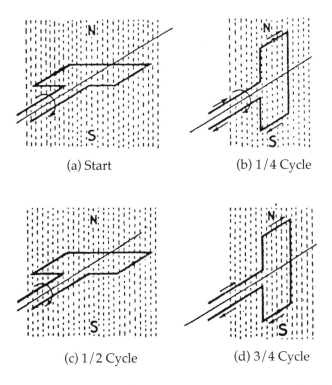

(a) Start   (b) 1/4 Cycle

(c) 1/2 Cycle   (d) 3/4 Cycle

Figure 6-12. Loop of wire rotating in magnetic field.

opposite to that during the first "half cycle." That is, the voltage will rise
in the opposite direction until a maximum's reached at "3/4 cycle" and
finally return to the original position where the voltage is zero. Thus as
the conductor passes under a pair of North and South poles, a cycle of
voltages takes place being first in one direction and then in the other.

Sine-Wave Voltages

Figure 6-13 shows these changes in voltage graphically; the
curved line represents the voltage at any instant by its vertical distance above or below the horizontal axis line. At the start the voltage
is zero when the conductor is moving parallel to the magnetic field.
Then the voltage rises to a positive maximum as the conductor passes
under the poles at "1/4 cycle," fails to zero, reverses, goes to a negative maximum as the conductor passes under the opposite pole and
finally returns to zero as the conductor completes the cycle and starts
on another one. The current in this circuit rises and falls with the volt-

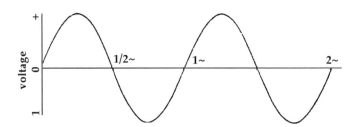

Figure 6-13. Curve showing voltage variations in wire loop rotating in magnetic field for two revolutions.

age; it flows alternately first one way and then the other, hence the name "alternating current." See also Figure 6-14.

The shape of this curve is known mathematically as a sine-wave.

Conductors in Series or Parallel

Suppose now another conductor is added diametrically opposite to the one already investigated. As it turns a complete revolution, it too will go through the same series of reactions as the original conductor. Furthermore, the direction of rotation and the relative positions between conductor and magnetic field will be identical to those of the first conductor so that an identical voltage (and current) will also be produced in the second conductor. The output of this machine has obviously been doubled, but the work necessary to revolve the two conductors will now be twice as great.

The two conductors, each producing a voltage, may be connected to independent outside circuits, or they may be connected together to one outside circuit as shown in Figure 6-15(a). The two conductors are in multiple and the electrical pressure or voltage—applied to the circuit will be the same as the voltage of only one conductor, but the current output will be the sum of the currents, in each conductor or twice that of one conductor. When connected as shown in Figure 5-15(b), the two conductors are in series and the voltage applied to the outside circuit will be the sum of the two voltages produced in each conductor, but the current output will be that of one conductor as the same current flows through both of them. Thus, conductors may be connected in parallel for greater current output or in series for greater voltage; in either case, the amount of mechanical work changed to electrical energy is the same.

The "generator" shown in Figure 6-16 is called a two-pole, single-

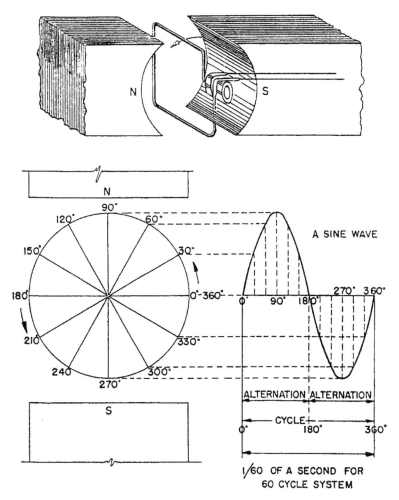

Figure 6-14. Sine wave alternating voltage/current.

phase, evolving-armature, alternating current generator. The magnetic field, the coils of wire and the iron core are called the "field" of the generator. The rotating conductor in which the voltage is produced is called the "armature."

If the voltage completes 60 cycles in one second, it is called a "60-cycle" voltage. The current that this voltage will cause to flow will be a 60-cycle current.

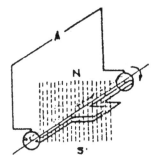

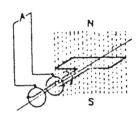

Figure 6-15a. Two conductors in parallel rotating in magnetic field.   **Figure 6-15b. Two conductors in series rotating in magnetic field.**

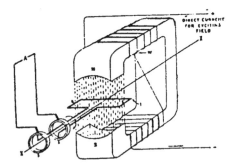

Figure 6-16. Simple single-phase revolving armature generator.

Two-Phase Generators

Figure 6-17(a) shows two single-phase generators whose armatures are mounted on one long shaft and must revolve together, always at right angles to each other. Then, when "phase 1" is in such a position that the voltage in it is a maximum, "phase 2" is in such a position that the voltage in it is zero. A quarter of a cycle later, phase 1 will be zero and phase 2 will have advanced to the position previously occupied by phase 1, and its voltage will be at a maximum. Thus phase 2 follows phase 1 and the voltage is always just a quarter of a cycle behind, due to the relative mechanical positions of the armature.

Figure 6-17(b) is a diagram of the relation of these voltages. It is seen that phase 2 repeats all the alternations of phase 1 a quarter of a cycle later.

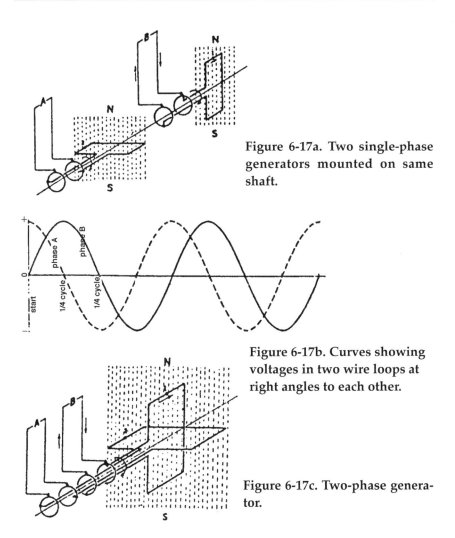

Figure 6-17a. Two single-phase generators mounted on same shaft.

Figure 6-17b. Curves showing voltages in two wire loops at right angles to each other.

Figure 6-17c. Two-phase generator.

Figure 6-17(c) shows how these two generators are put together and rotate through the same magnetic field on the same shaft, making a two-phase generator, each external circuit being separate just as though there were two single phase machines. The two separate voltages produced are displaced electrically 90° from each other.

Three-Phase Generators

In a three-phase generator, three single-phase windings are combined on a single shaft and rotate in the same magnetic field as shown

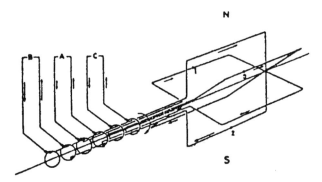

Figure 6-18a. Three single-phase generators mounted on same shaft.

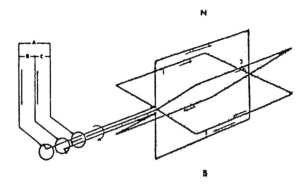

Figure 6-18b. Three-phase generator.

in Figure 6-18(a) and (b). Each end of each winding can be brought out to a slip ring and an external circuit. In the three-phase generator the voltage in each phase alter nates exactly a third of a cycle after the one ahead of it due to the mechanical arrangement of the machine. When the voltage in phase 1 is approaching a positive maximum, that in phase 2 is at a negative maximum and the voltage in phase 3 is falling.

As a rule, however, each end of each phase winding is not brought out to a separate slip ring in a three-phase machine, but the three phases are all connected together in some arrangement as illustrated in Figure 6-18(b). This makes only three leads necessary for a three-phase wind-ing, each lead serving two phases. Then each pair of wires acts like a single phase circuit, substantially independent of the other phases.

PHASE WINDINGS

Machines could be built for almost any number of phases, but this requires more expensive construction, particularly in the number of slip rings, and, as will be seen later, there is little advantage in using more than three phases. Consequently, machines having more than three phases are seldom manufactured except for certain machines which convert AC to DC and often make use of six phases.

In this discussion, it will be noted that the term phase really means portion. For example, the output of a two-phase machine comes in two portions, that of a three-phase machine in three portions. It will also be noted that as more conductors are added in the machines, better use is made of the space around the armature. These conductors are usually wound around an iron core and the whole assembly rotated in the magnetic field.

To make still better use of the space around the armature, the winding for a phase may consist of several turns, instead of the single loop shown in the previous discussion. However, the voltage generated in each turn of the winding will be slightly out of phase with the others and so the spread of the winding around the armature must be confined to a relatively narrow section. The voltages produced in the different turns of the winding will then add up to zero or some small value as well as cause the currents produced by each voltage to buck each other and dissipate the electrical energy in the form of heat.

MULTIPOLAR MACHINES

A generator may have a single pair of north and south magnetic poles; or it may have more than one pair. The number of poles is always even because for every north pole there is a corresponding south pole.

In Figure 6-19(a) is shown diagrammatically a four-pole, single phase, revolving armature generator. It may be seen from this diagram that any conductor passes first under a north and then under a south pole, but that in one revolution of the shaft, this conductor has passed by four poles, that is, two pairs of poles, giving two complete cycles for one revolution. If there were six poles in the machine, there would be three cycles, one for each pair of poles, in each revolution.

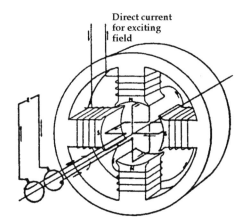

Direct current for exciting field

Figure 6-19a. Simple four-pole, single-phase revolving armature generator.

FREQUENCY

If a generator has two pairs of poles and runs at 1800 revolutions per minute, it will make two times 1800 or 3600 cycles per minute. The frequency of a machine is always measured in cycles per second, so that the cycles per minute must be divided by 60 to get frequency. In the example given above, the frequency will be 3600 divided by 60, or 60 cycles persecond. Thus,

$$\text{Frequency (cycles per second)} = \frac{\text{RPM}}{60} \times \text{number of pairs of poles}$$

The frequency most in use in the United States is 60 cycles per second (60 Hertz).

A generator designed to be driven by a steam engine, oil engine, large water turbine or other slowly revolving prime mover must run slowly if it is to be directly connected; consequently it must have many poles, sometimes 30 or more. On the other hand, generators that are to be connected to steam turbines often have only two or four poles because they must run at high speeds.

REVOLVING-FIELD TYPE MACHINES

As mentioned earlier, a stationary conductor which has a magnetic field sweeping across it is cutting the magnetic field just the same as though the conductor were moving across the magnetic field. Such a machine, in which the armature is stationary and in which the steady magnetic field revolves is known as a revolving-field machine. In all the diagrams presented thus far, the poles making up the magnetic fields have been stationary on the frame of the machine and the armature in which the voltages are produced has moved. Alternating current generators, however, usually have revolving fields because it means that only two slip rings need be used. These two carry the low voltage "exciting" current to the field, while for a three-phase machine at least three slip rings would be required for the armature current which is often generated at a high voltage. High voltage windings that remain stationary can be insulated more effectively than when they must also resist the additional mechanical stresses of rotation. The result is the same whether the field or the armature revolves, since in either case, the conductors cut across the magnetic field. Machines are built whichever way gives the best construction. Figure 6-19(b) shows a diagram of a single-phase revolving-field generator.

DIRECT-CURRENT GENERATORS

Figure 6-20 shows diagrammatically a direct-current generator which corresponds to the single-phase generator shown in Figure 6-11. In the DC machine, a mechanical device called a "commutator" is used to reverse the connections to the revolving conductors in the generators at just the instant the current in them is reversing. Each end of the loop of wire is connected to one segment of commutator, each segment being insulated from the other. The sliding contacts or brushes are so placed

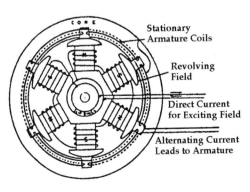

Stationary
Armature Coils

Revolving
Field

Direct Current
for Exciting Field

Alternating Current
Leads to Armature

Figure 6-19b. Single-phase generator with six poles in a revolving field.

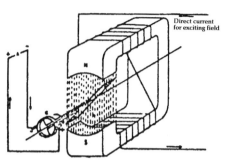

Figure 6-20. Simple direct-current generator. Note the commutator used to keep the current in the associated circuit flowing continuously in the same direction.

that the current always flows from the loop into one and from the other into the loop regardless of which relative direction the current in the loop itself flows. In this way the brushes carry current flowing in one direction only. The brushes or terminals are designated by the terms "positive and negative" simply to distinguish the direction in which the current is assumed to flow.

Exciters

Exciters are small direct-current generators usually mounted on the same shaft as the generator. Their purpose is to provide direct current to the electromagnets which create the magnetic field cut by the conductors in which electricity is generated. Brushes provide a contact with rings that make the connection between the output of the exciters and the rotating electromagnetic poles of the generator.

Generator Construction

Relatively small size generators, approximately 1000 kW or less, may be driven by internal combustion or steam reciprocating engines. Here, the mass of the rotor may not be sufficient to maintain the momentum required to produce smooth and continuous voltage in the generator. To accomplish this, a separate fly-wheel may be provided in those instances where the prime mover may not be equipped with one, or where its fly-wheel may not be sufficient.

Large generators are almost always driven by steam turbines and the mass of their rotors makes fly-wheels unnecessary. The limitations of the strength of available materials together with the speed of rotation determines the diameter of the rotors: for 1800 rpm, 4-pole, generators this limit is about six feet; for 3600 RPM, 2-pole generators, the limit is about 3.5 feet. With these limiting dimensions, the only way to increase the output of the generators is to increase the core length. Safety considerations also impose limits to core lengths. For example, for an 1800 RPM machine with a core diameter of six feet and length of 20 feet, the peripheral speed is in the nature of 35,000 feet per minute and every pound of material in the periphery experiences a centrifugal force of over two tons.

Most turbine generators are of the horizontal-shaft type. Waterwheel generators, however, are usually of the vertical-shaft type because this provides a more satisfactory mechanical arrangement for the waterwheel. The speed of waterwheel generators varies widely with the head of water under which the prime movers operate.

Losses

In every machine, there is a difference between the energy input and the energy output; the differences are due to losses that occur for several reasons. These include the mechanical losses caused by friction of the rotating parts and that caused by the interaction between the rotor and the air in the gap between the rotor and the stator called windage. Electrical losses include those in the steel core of the stator caused by the alternating magnetic field's action on its molecules, on those caused by current flowing in the conductors of both the stator and rotor (I^2R), and so-called stray losses caused by eddy currents generated in the steel cores and other metallic parts of the generator cut by the moving magnetic fields. All of these losses manifest themselves in the form of heat, which requires that the machines be cooled by ventilation and other means.

Cooling Ventilation

Cooling of the generators is generally accomplished by forcing air through passages in the stator laminations and by blowing air over the stator coil ends. Fans mounted on the rotor shaft blows air directed over the rotor coils. The whole assembly is cooled by the air flowing through the space between poles and by the movement of the air in the air gap; in some instances, axial ducts in the rotor aid in the circulation of air.

Open type machines may be cooled by fans, but air pressures will not be built up within the machine, and cooling will depend on the inertia imparted to the air by the fans and moving parts of the machine.

Enclosed type machines have solid covers forming a closed space into which fans on the rotor or external fans force air and build up a pressure. For large machines, the same air may be circulated through air coolers which may be part of the generator assembly, or may be circulated outside the machine through fire-tube water coolers.

In closed type machines, the air may be replaced by hydrogen or helium gas.

The primary reason for using hydrogen is the considerable reduction in windage loss to about one-tenth of that of air and to provide better cooling so that the rating of the machine may be increased some 20 percent or more. Other advantages of hydrogen are the reduction of oxidation of insulation and metallic parts and the reduction of windage noise.

Extra equipment needed for hydrogen cooling include: an explosion proof housing; means of controlling admission and purity of the gas and of exhausting it without permitting explosive mixtures to form; seals to retain the hydrogen where the shaft passes through the housing; and, means of treating lubricating oil to maintain its purity. These additional costs are more than offset by the increased efficiency and the overall economic benefits derived from the better cooling of hydrogen.

Further, the greater cooling that can be obtained with hydrogen at pressures well above atmospheric make possible the carrying of overloads by increasing the hydrogen pressures during periods of overload.

Helium is the next best gas to hydrogen. Its density is about half again that of hydrogen while its heat transfer capability is about 80 percent that of hydrogen, values still better than air. While its cost is much greater than that of hydrogen, its nonexplosive nature eliminates some of the costs associated with hydrogen and the perception of an increase in safety justify the use of this gas.

Voltage Regulation

To keep the output voltage of a generator approximately constant under changing loads, provision is made to vary the d-c voltage provided by the exciter that determines the strength of the magnetic field cutting the generator conductors. The exciter output may be varied by a simple rheostat controlled by a contact-making-voltmeter for smaller units, or by electronic devices providing a much more refined control, especially under contingency conditions of faults or rapid fluctuations in load. See Chapter 5, Figure 5-20.

Frequency Control

A constant frequency, usually of 60 cycles per second, is important for the operation of many utilization devices and is increasingly demanded by consumers. While the frequency is determined by the speed of rotation, and this is controlled by the prime movers, in many instances, especially for turbine driven generators, additional control devices may be employed. Self synchronizing (selsyn) devices previously described for control of turbine speed, may also be employed for control of generator speed. In some cases, these may be interconnected so that the speeds of both turbine and generator may be coordinated. See Chapter 4, Figure 4-21.

Turning Gear

For large machines, the mass of the rotor may be large enough to cause the bending of the shaft when the generator is at rest or coming to a rest. Turning gear is employed in gradually raising the generator to speed before supplying load, just as for the associated turbine. Likewise, when load is removed, the turning gear may be used in gradually reducing the generator speed to zero. In some instances, one turning gear is used for controlling both the turbine and the generator, in starting up and shutting down.

Associated Electrical Facilities

Having described how the energy contained in fuel is converted to electrical energy, there remains to describe the means by which this energy is to leave the generating station and to enter the delivery systems on its journey to its ultimate consumers. These facilities can be considered as the output part of the generator.

Main Electrical Connections

These are determined by a number of factors, the principal ones being:

1. Size, number, and voltage of the generators.

2. Characteristics of the delivery systems to be supplied.

3. Reliability requirements and their relation to the entire electrical supply system.

4.   Operating practices.

5.   Maintenance practices.

Overriding considerations of prime importance in all of these factors are those of safety, public and environmental requirements, and economics, generally in that order.

Bus Arrangement
A bus is a convenient means by which the conductors from generators may be connected together, and connected to conductors which may be part of the delivery systems.

Single Bus
This is the simplest and most economical arrangement. It is generally confined to small, single unit stations where service interruptions can be tolerated. In case of failure in any of its components, or the need for their maintenance, the entire station may be de-energized; while maintenance may be accomplished with the bus energized, it is an unsafe and undesirable practice.

Transfer (or Spare) Bus
This is a variation of the single-bus arrangement, but one which allows any of the components to be de-energized for maintenance, but which may cause the entire station to shut down in event of fault in any of the components. It is fairly inexpensive and is generally confined to small stations.

Double Bus—Single Breaker
This arrangement reduces the duration of outages due to bus failure and facilitates bus maintenance. Feeder breaker failure or maintenance will result in outages to the feeder. The bus-tie breaker permits the transfer of feeders from one bus to the other without service interruption. While still relatively inexpensive, it is generally confined to small and medium sized stations.

Double Bus—Double Breaker
Although more expensive, this arrangement reduces the duration of outages on any component and permits maintenance without

service interruption. A circuit breaker failure, however, will cause the entire station to shut down until the faulty unit is isolated. Additional back-up breakers may be installed reducing bus outages due to feeder breaker failure and permitting the maintenance of any of the components without affecting other components. It can be economically justified only in large stations or where service continuity is of prime importance.

Group-Bus

This is a modification of the double bus-double breaker arrangement, but less expensive. It is particularly applicable where a large number of feeders may be involved.

Bus Sectionalization

Bus sectionalization is an important feature in main electric arrangements and may be used with any of the arrangements described above. Three sectionalizing arrangements are commonly used and are known as: straight bus; ring bus; and star or wye bus.

Straight Bus

In this arrangement the buses are divided into sections between which are circuit breakers that may be operated normally open or closed depending on the type system being supplied.

Ring Bus

In this arrangement, connections between the bus sections are maintained, even with the loss of one section and facilitates transfer of generators between sections. Other double bus arrangements may be modified into ring buses by tying the buses together with a circuit breaker.

Star or Wye Bus

This scheme consists essentially of sectionalized main and tie bus arrangement in which both buses are in normal operation; it makes any of the generators automatically available to any of the main buses. Various modifications are also used depending on the degree of service reliability desired.

Other Bus Arrangements

In some large power systems, the generating voltages, usually no higher than 30,000 volts because of insulation limitations, needs to be raised to accommodate supply systems. In such instances, transformers are employed to serve this purpose.

Since the operation of the circuit breakers in any bus arrangement is of vital importance to the safe operation of the electric generating station, the power necessary for their operation is often supplied by batteries which are under continual charge. The independent source assures the operation of the breakers under contingency conditions that may affect the rest of the generating station. Auxiliary lighting at critical locations may also be supplied from the batteries.

Mechanical Considerations

Precautions are taken in the design and construction of power plants to prevent failures of the several components and, if failure does occur, to limit the effects of the failure. In addition to the automatic operation of the circuit breakers to de-energize the separate components under conditions of fault or incipient fault as indicated by overload and high temperature indications, other measures include their physical separation and isolation as much as practical.

Where more than one generator exists in a power plant, each generator is separated from the other by a vertical fire and explosion-resistant wall or barrier.

As mentioned earlier, alternating current generators may consist of three separate circuits, or phases, usually with one end of each circuit connected together in a star or wye connection, with the common connection normally grounded. Each phase is separated from the others horizontally, each phase bus being on a separate floor or level, separated by fire and explosion proof floors or barriers.

Circuit breakers and transformers involved also are each isolated within protective fire and explosion proof barriers.

Generators, buses and other equipment may be protected from fire and flame by automatically operated carbon dioxide (or other chemical) sprays, actuated either by temperature or overcurrent protective devices, or both.

Grounding

Further protection is obtained by the application of grounds, that is, connections to earth. These may be simple bars driven into the earth, a mesh of wires buried in the earth, connection to buried metallic structures, or a combination of some or all of these.

Grounding at the stations is of the greatest importance and has the following principal functions:

1. Grounding of the neutral or common point of star or wye connected phases of a generator or a bank of transformers reduces the strain on insulation, and permits more sensitive settings of protective relays and of protective surge arresters.

2. Provides a discharge path for surge (lightning) arresters, gaps, and other similar devices.

3. Insures that non current carrying parts, such as equipment frames, are always safely at electrical ground or zero values should equipment insulation fail.

4. Provides a positive means of discharging and holding at zero electrical values for equipment (and feeders) before and while proceeding with maintenance on them.

Auxiliary Power

In modern power plants, many of the essential pumps, fans, conveyors, heaters, exciters, turning gear, and station lighting are electrically operated. To provide a secure source of electrical power, feeders from two or more separate sources are supplied. These sources include separate buses from two or more generators, a separate house generator powered by steam turbine, reciprocating steam engine, diesel or gasoline engines, motor-generator sets from stand-by batteries, and from external distribution and transmission systems directly or through separate banks of transformers. The separate house generators may be kept solely as stand-by to be used to start the station from a "cold" start, or they may be kept spinning for emergency purposes. At least two such sources of auxiliary power are included in small and medium size fossil fuel plants, and three or more are used to support operations in nuclear plants.

Circuit arrangements often call for individual motors to be supplied from individual feeders controlled by switches at some control point or board. In other instances, a unit system arrangement is provided whereby a number of associated motors (e.g., boiler feed pump, circulating pump, forced air fan, induced draft fan, etc., of a unit) is supplied from a common feeder; care is taken that duplicate or reserve facilities associated with a unit are not connected to the same feeder.

Special battery supply for protective devices and for the operation of circuit breakers is usually provided.

Chapter 7

Operation and Maintenance

T he coordination of all the equipment and processes involved in producing electrical energy from fuel or water require the continuous acquisition and evaluation of data from which appropriate action by the operators may be taken to achieve a safe, efficient and reliable operation of the generating plant. In many older plants much of the operation is accomplished manually by trained, skilled personnel. In newer plants, much of the data acquisition and operation of required actions are performed automatically by mechanically or electrically actuated devices. In the newest plants, essentially all of the data acquisition and operation of required actions are performed with the use of computers. This not only reduces the amount of personnel required to operate the plant but permits the rapid, detailed and accurate correlation of operating conditions which in turn results in quicker response to changing conditions imposed on the several elements comprising the generating plant.

Quicker response and more frequent adjustment of controls that are made possible by modern electronic equipment and computers result in higher operating efficiencies and reduced cost to both the utility and the customer. Where feasible, older plants are having their control systems replaced and converted to the latest electronic equipment.

Although "redundancy" (duplicate systems or more to back up critical operations) is included in the control and protective devices, as well as "fail safe" features in their operation, personal supervision is usually provided as the final authority to which the operation of the plant is entrusted.

INSTRUMENTS

All of the data from the "boiler room" and the "turbine room," the mechanical and electrical elements of the plant, are brought from instruments and sensing devices to a central control room (sometimes referred to as the control board). The data are shown on instruments called meters and may be of two types: indicating, which shows the moment to moment conditions; and recording, which accumulates the data over a period of time constituting a history of the particular equipment or procedure. In modern plants or those that have been converted to burner management systems, the computer is used to log and display such data. Although some of the processes involve mechanical properties (e.g., steam flow, water flow, temperatures, etc.) the mechanical quantities are converted by devices called "transducers" into electrical quantities that can be more readily transmitted by wires or fiber optics (instead of piping) to the control board instrumentation or computer.

Steam Boiler

For the efficient operation of the boiler (including the furnace) at any specific time, initiate necessary adjustment of equipment to take care of changes in load and other conditions. The instruments furnishing data for each unit include:

1. Steam flow meter to determine the boiler output.

2. Draft gauges for stack pressure, furnace draft: incoming and outgoing gas.

3. Thermometers for measuring temperatures of flue gases leaving the boiler.

4. Air flow meters supplying air for combustion.

5. Temperature indicators to measure the steam temperature leaving the superheater.

6. Recording steam pressure gauge.

7. Fuel combustion rate: weight of coal, volume of oil or gas per minute.

8.   Television camera for observation of flame condition in furnace (in modern installations).

9.   Gas analysis for smoke control.

Nuclear Reactor

In addition to the instruments mentioned above in items 1, 5 and 6, there are also instruments to measure:

1.   Radioactivity of the core.

2.   Position of the control rods.

3.   Temperature within the reactor.

4.   Temperature on incoming water or steam, or both, and of outgoing steam.

5.   Radioactivity, if any, of escaping gas.

6.   Radioactivity, if any, of ambient surrounding the reactor.

Prime Movers

These will vary with the type of engine, the only instrument common to all is a:

1.   Tachometer for registering speeds of rotation in rpm. For internal combustion engines (similar to cars) there are also:

2.   Temperature of coolant.

3.   Lubricating oil level and pressure.

4.   Alternator output for ignition and for charging of battery, if any.

5.   Fuel gauge, indicating rate of consumption.

6.   Air flow meter for air supplied for combustion.

7.   Exhaust CO and CO_2 content.

8. Thermometers for measuring inlet and outlet temperatures where external cooling devices for coolants are used.

9. Water flow gauge for measuring water input and velocity for hydroelectric installations.

For reciprocating steam engines, in addition to the tachometer, there are also:

1. Steam pressure gauges for inlet and exhaust steam.

2. Thermometers for measuring temperatures of inlet and outlet steam.

3. Thermometer for measuring coolant temperature, where used.

4. Thermometer for measuring temperature of lubricating oil.

5. Thermometers for measuring inlet and outlet temperatures for coolant, where external cooling devices are used.

For steam turbines, in addition to the tachometer, there are:

1. Steam pressure gauges for inlet steam and exhaust steam at each of the bleed points.

2. Thermometers for measuring temperatures of inlet steam and exhaust steam at each of the bleed points.

3. Thermometers for measuring temperatures of lubricating oil at main bearings.

4. Device for measuring eccentricity of turbine rotor shaft.

5. Pressure gauge for measuring pressure of turbine steam inlet to condenser.

6. Thermometers for measuring temperatures of turbine steam inlet to condenser and of water-steam outlet from condenser.

7. Control for turning gear.

For electric generators, in addition to the tachometer, there are also:

1. Thermometers to measure the temperatures of the unit and, where it is of the enclosed type, temperatures of input and output air or gas.

2. Pressure gauges for measuring pressures of input and output air or gas, where external cooler is used, and for determining leaks.

3. Thermometer for measuring temperature of input air or gas where external cooler is used.

4. Thermometer to measure temperature of lubricating oil in main bearing.

5. Device for measuring eccentricity of rotor.

6. Control for turning gear.

7. Voltmeters (AC) to determine output voltages on each phase of the generator, both indicating and recording, and a (DC) for measuring exciter voltage.

8. Ammeters to determine current output on each phase of the generator, both indicating and recording, and a (DC) for measuring exciter current.

9. Wattmeters, both indicating and recording, for determining power output of generator.

10. Watthour meters, both indicating and recording, for determining energy output of generator.

11. Frequency meter, both indicating and recording.

12. Power factor meter, both indicating and recording.

13. Reactive volt-ampere meters, both indicating and recording.

14. Reactive volt-ampere hour meters, both indicating and recording.

15. Synchroscope for synchronizing units to line before connecting.

16. Synchronizing control.

17. Operating controls and position indicators for each circuit breakers.

18. Voltmeters to measure bus voltage for outgoing feeders.

19. Ammeters to measure current in each outgoing feeder.

Other Instrumentation

1.   Thermometers, both indicating and recording, for measuring outdoor temperature.

2.   Barometer, both indicating and recording, of outdoor atmospheric pressure.
3.   Wind direction indicator, both indicating and recording.

4.   Light meter for determining natural light, both indicating and recording.

5.   Rain gauge for determining rainfall.

Alarms and Indicators

In addition to the meters, there are also alarm sound and light signals to call attention to abnormal conditions, including fire alarms for several locations in the plant, in the coal pile and hoppers and fuel containers, both for oil and gas.

Indicators of the positions and operations of the many protective devices installed at all of the elements of the generating plant.

Communications

At each control board may be found telephones, both for internal communication and for connection to outside private and commercial systems, short wave radio for communication with outside forces, telegraphic signals to various centers of the station and, with special telephones, to public fire and police departments.

This is an awesome display of meters, controls, and other indicators which, no doubt, would leave the layman astonished and impressed, but which is taken in stride by the trained operators and other personnel. Not only do the visual and sound alarms call attention to those areas requiring appropriate action, but periodic rehearsals help the operators to meet contingencies quickly and accurately

START UP AND SHUT DOWN

Perhaps the most important procedure in operating a power plant is the starting up of the several units within the plant and their shut-down, both under normal and emergency conditions. The methods and procedures for each of the several type units are generally different, and will be dealt with separately.

Hydro Plants

This type plant presents the fewest problems. As the generator rotor is usually coupled directly to the water wheel, because of the large mass involved, it will take a little time to start up and until the combination is brought up to speed before load can be applied to the generator. Similarly, for a normal shut down after the load is gradually disconnected from the generator, the water supply is gradually turned off until the wheel and the generator rotor came to a halt. As the speed of revolution is relatively low, in the nature of 600 rpm (as compared to 1800 and 3600 for steam turbines), the inertia to be overcome is proportionally also less and time necessary for start up and normal shut down is therefore also relatively low.

Should the electric load be suddenly disconnected (in the event of fault), the tendency for the water wheel and generator rotor to accelerate beyond normal speed is great and takes place, and will set up stresses in the rotating mass. The unit, however, is designed to accommodate these temporary stresses safely.

Internal Combustion Engines

In gas, gasoline and oil fueled engines, start up is simple, just as starting of an automobile. Often they nay have to be warmed up before load can be applied, the period of time may vary depending on the ambient and other causes, but generally is not significant.

In most cases, the generator is attached directly to the engine, and little or no difficulty is experienced as load is applied gradually to the generator. In some larger diesel installations, the generator may be connected to the engine through a clutch arrangement after the engine is started and warmed up.

Shut down is normally accomplished by gradually decreasing the fuel supply as the load on the generator is decreased. In the event when the electrical load is suddenly disconnected (as in a fault), the engine will tend to speed up, but the fuel supply mechanism which keeps the engine rotating at a constant speed, will operate rapidly to bring the engine back to normal speed, and bring it to a halt in the normal fashion. The relatively slow normal operating speeds for these type engines, usually less than 1000 rpm, present little or no difficulty during the shut down operations.

Reciprocating Steam Engines

In these installations, the generation of steam must be considered. From a "cold" condition, several hours may elapse before steam in sufficient quantity and pressure is generated to operate the associated engine. The type of fuel used, whether lump coal, pulverized coal, oil or gas, may affect the rate at which steam is produced, but a minimum time is usually maintained to enable the' several parts of the furnace and boiler, especially the furnace walls, to expand at a sufficiently slow rate to prevent serious damage from ensuing.

The steam engine is theoretically able to operate at once, picking up load as soon as it receives steam. In practice, however, some time elapses in which the engine warms up.

Electrical generators in these instances are generally connected by a belt or sometimes by a series of linkages. Steam in the cylinder acting as a "cushion," and slipping of the belt, permit the generator to pick up load smoothly. Decreasing the supply of steam while maintaining speed accommodate the decrease in load on the generator gradually until the units may be shut down. The boiler and furnace, however, are lowered in temperature gradually over a period of several hours to prevent possible damage described for start up.

Should load be disconnected suddenly from the generator, it will tend to speed up, but slippage of the belt and play in the linkages will tend to dampen the effect on the steam engine. The steam engine will, however, tend to speed up and the steam pressure within the boiler may tend to increase as the demand for the amount of steam produced will diminish. If this pressure rises to values deemed undesirable, the old fashioned safety valve will operate to let off steam at a rapid rate until its pressure is reduced to safe values.

Steam Turbines

The same three elements exist in these installations as in the reciprocating steam engine. In general, the conditions that affect the steam supply by the furnace-boiler combination also apply. The turbine and generators, usually much larger in size and mass, and operating at much higher speeds (1800 and 3600 rpm), however, encounter some additional problems. The turbine and generator are usually connected into one unit.

Because of the mass of the turbine rotor, a longer time will be required to bring it up to the necessary speed; moreover, the shaft of the rotor may actually be bowed a small amount, but which could cause damage to the turbine because of the very small clearance between the rotor and stator. A turning gear is provided to turn the rotor very slowly and gradually accelerating it to its rated speed, sometimes aided at the higher speed by the introduction of steam. In shutting down, the reverse procedure is generally followed, with a gradual diminishing of steam input and the turning gear slowing it to a stop in the later stages.

(A procedure similar to that for the turbine is followed in starting and stopping the electrical generator, and essentially for the same reasons.)

In the event of fault or other condition which would suddenly disconnect the load from the generator, both the generator and turbine rotors would tend to speed up, throttle adjustments would decrease the steam input. The turbine and generator are designed to support the temporary speed increase while they would gradually slow down until the turning gears enter the operation. The excess steam produced in the boilers during this interim period would operate the safety valves, just as described above for the reciprocating steam engine installations.

Nuclear Reactor

These are found in large installations employing steam turbines (operating at lower steam pressures of about 800 pounds per square inch) and large high speed (1800 and 3600 rpm) generators. The same procedures apply to the turbine and generator as described above. The procedures for the nuclear reactor are quite different than those for fossil fuel steam generators (furnace-boiler).

In these installations, steam is generated from the heat resulting from the radioactivity taking place between the uranium fuel contained in rods. The reaction takes place as the control rods, made of graphite or other material that stops the flow of neutrons, are raised slowly causing the interaction between fuel rods to begin generating heat producing steam. The reverse action stops the generation of heat and steam. Once started, however, the radiation that occurs as part of the process causes surrounding supports, containers, etc., to become radioactive permanently. Hence, the unit cannot be shut down or "deactivated" in a short period of time, unlike that for fossil fuel installations.

"Scramming"

In boilers fired by pulverized coal, oil or gas, the flames or source of heat can be turned off immediately by turning off the fuel supply to the jet burners. In a nuclear reactor, besides the set of control rods, there is provided another set of safety rods which are fitted with quick-acting latching devices and so interlocked in the control gear that they can be inserted instantly, halting the flow of neutrons; this rapid emergency action is called "scramming."

To supply the needs of the nuclear plant in event of a complete shut down, gasoline or diesel fueled auxiliary generators are installed at such power plants, and these are further supplemented by transmission supply from other sources distant from the nuclear plant.

EFFICIENT USE OF GENERATION

Obviously there is a great difference in the efficiency of operation for various units because of size, age, fuel costs and labor, and many other tangible and intangible considerations, such as local ordinances, national policies, future requirements, seasonal ambient variations, etc. As systems grow and expand, even within an individual power plant units of varying efficiencies may exist to serve loads that differ from hour to hour, day to day, month to month during the year. Prior to deregulation the selection of units was based on~the efficiency of the units within the utility area and the price of energy available thru interconnection. A typical graph of hour to hour demand over a twenty four hour period, called a load curve, is shown in Figure 7-1.

To illustrate selection of units, unit I would be the most economically efficient of available generating units (typically a large steam powered generator) and is used for the full twenty four hour period to serve base load. Unit II, would be the next most efficient unit (a smaller perhaps older unit) put in service for this increment of load. The remaining increments would be served by more costly units that are easy to start and stop over short periods, such as a combustion turbine or internal combustion engine powered.

Deregulation has dramatically changed this picture with the divestiture of generation by power delivery utilities, the creation of ISO's (independent system operators) and a market-driven dispatch of generation by regions, including the effect of transmission constraints. A utility must now work through the ISO to either buy or sell power (if the utility still has generation) on an hour by hour bidding system. The utility may also bid or engage in "day ahead bids" (futures) as well as periodic firm capacity agreements.

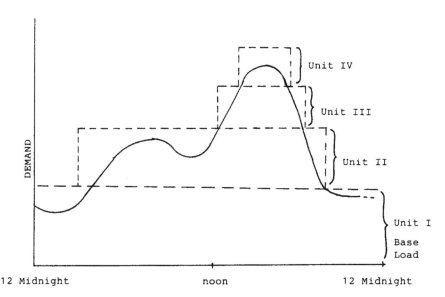

Figure 7-1. Assignment of units according to their economic efficiencies in supplying varying daily load demands.

Units are committed not always by the most efficient but by what the seller is offering at a price he determines at the time and what the buyer is willing to pay for, as in stock market operation. The impact on customers served by the utility and on the utilities fmancial viability can be illustrated by the California experience. Aside from an economic impact, such operations may also affect the system reliability or the quality of service to the ultimate customer.

RELIABILITY-SPINNING RESERVE

To insure an uninterrupted supply of electricity, consideration is given to the possibility of the failure of one or more generating units within either a generating plant or comprising an electric grid tying together several generating plants within a geographic area (the units may pertain to one or more utilities). Depending on the reliability desired, generators whose output is equal to the largest single generator connected in the system are kept in readiness to pick up the electrical load should that largest unit fail or be out of service. In some instances, where greater reliability is desired, provision is made to maintain service should two, and in some cases even three, of the largest units become unavailable.

Although the illustration above includes four different types of generating units, most plants may have only one or two different types. Where the reserve units are steam turbine driven, for the reasons explained earlier, they are kept in operation, known as "spinning reserve."

In a generating grid or power pool, the capacity of some part of the reserve units may be substituted by a firm transmission source constituting part of the grid.

PARALLELING UNITS-SYNCHRONIZATION

Great care must be exercised in connecting a generator directly to a bus or circuit to which another generator is connected; that is, when generators are connected in parallel to share the supplying of electricity.

Referring to Figure 6-10 in Chapter 6, the water analogy for both

DC and AC systems applies. In the DC system, if a second driving pump and piping were to be connected to the common nozzle driving the water wheel (to which the water motor is attached), the pressure developed by the second pump must be the same as the first or part of the pressure in the first would act to hold back the water in the first system. To be effective, the pressures in both systems must be the same.

In the AC system, a second system paralleling the first must not only develop the same pressure, but the pistons of the driving pumps in both systems must also move at the same speed and in the same relative position to (in step with) each other, that is in synchronism, otherwise one system would work against the other to the point where the pistons would come to a halt as the pressures of the first system would counterbalance those of the second system.

Referring to Figure 6-13 in Chapter 6, the sine wave curve delineating the voltage (or pressure) produced in one system must have the same magnitude, move at the same speed, and be in step with each other, to be completely effective.

This process is known as synchronization, that is, the AC generators involved are synchronized so that the voltages produced by each unit are exactly the same. This may be accomplished automatically by a device known as a synchroscope, a device which measures the difference in voltage produced by each of the generators while separate from each other, and indicates when the voltage difference is zero and the time when switches or circuit breakers may be closed connecting the two units together electrically.

CONTROLS AND PROTECTIVE DEVICES

Many of the control and protective functions required for the operation of a generating station are performed by automatic devices, which in turn are supervised by instruments and alarms at the control board indicating their proper functioning or malfunctioning. In a few of the smaller and older stations, some of the functions may be done manually. Generally, however, as further insurance, operators are assigned around the clock to provide human intervention as a further precaution against unforeseen contingencies.

Some newer and smaller generating stations, principally hydro plants situated in remote locations, are fully automated and are visited only periodically for maintenance purposes. Other plants may be operated semi-automatically from other stations via telephone and radio communication signals.

Many of the control and protective devices are comprised of magnets, electromagnets springs, bellows, linkages, and other mechanical components which operate to connect or disconnect auxiliary motor driven or magnetic coil activated devices to accomplish the desired procedures. More recently, these devices are electronically activated that permit more rapid, reliable, responses as well as a reduction in maintenance costs. Many of the former exist and will exist for a long time; where economically feasible, however, they are replaced with electronic devices. In some of the larger more important plants, where greater reliability is desired, duplicate or multiple installations of such devices, as back up, are made; moreover, they may be designed to "fail safe," that is, the failure of the device and back up will deactivate the particular element which they are to protect. These observations are generally applicable to fossil fuel stations.

Nuclear reactor plants are almost universally equipped with electronically activated control and protective devices, with a redundancy (that is, more than one back up system, based on alternate systems of detection and operation, if practical) of such devices; these are also designed to fail safe. Moreover, modern computers permit the programming of normal and contingency operation of the plants that make possible not only greater reliability but also more efficient performance of the generating station.

MAINTENANCE

Maintenance of the several elements of a generating plant fall under three broad classifications: normal, emergency and preventive

Without going into extreme detail, normal maintenance includes such items as removing slag and clinkers from furnace walls and grates, sludge and scale from boiler tubes by blowdown and by mechanical scrapers, soot from boiler walls and stacks, soot and cinders from precipitators, replenishment of feedwater treatment materials, dressing or replacement of brushes on motors and generators, lubrication and replacement of lubricating oils, checking of adjustments on governors, regulators, relays and other instruments and alarm devices, routine inspections and checks on other equipment.

Emergency maintenance includes such items as repair of cracked furnace walls, cracked and broken boiler tubes, broken turbine blades, burned out or damaged motors, loose or broken electrical connections, and repair or replacement of any element that may affect the safety of personnel, and those that may cause the actual or imminent shut down of the plant, or seriously reduce the capacity of the plant monitoring by devices at the control boards and by alarm signals give indications of possible type and location of trouble and procedures for trouble shooting are continually updated.

Preventive maintenance usually calls for the "wholesale" replacement of equipment or parts of equipment throughout the plant before their time for normal repair or replacement becomes necessary. These may include such items as fans and pumps, motors, relays and protective devices, and other critical items, including such other items as station lighting, painting, etc.

Such maintenance activities for hydro and fossil fuel plants are carried out under safe procedures, with proper tools and equipment, by trained workmen under experienced supervision. Where practical, except in certain cases of emergency, equipment is taken out of service while it, or parts of it, are being maintained. The skills employed are those that may be found in many other industrial, manufacturing and commercial undertakings and include carpenters, mechanics, welders, electricians, test technicians, painters, and other similar crafts.

Nuclear Plants

In nuclear plants, maintenance requirements are generally more stringent as some of the equipment is radioactive, and the risk of exposure to undesirable amounts of radiation is always present. In these cases, special equipment which permits the handling of radioactive items is built into the unit as a permanent installation, as mentioned earlier, many of the items of equipment and devices are installed in duplicate (or even more in some cases) and, hence, provide the ability to take out of service for maintenance many of the components, even though some may be radioactive, without affecting the operation of the generating plant.

The people handling such equipment are dressed in radioactive resistant clothes, wear devices which signal the approach or existence of radioactivity, their time working in such ambients is restricted, they undergo decontamination procedures at the end of their work, and receive periodic medical examinations to discover any impairment to their health because of the nature of their work. There are, however, many other items within the nuclear plant that are not subject to radiation, and these are maintained in much the same manner as similar items in fossil fuel plants.

Chapter 8

Environment
And Conservation

From the planning stage right through the continuing operation of a generating station, the impact on environment and conservation are considerations uppermost in the minds of management. These considerations are factored into the economic ones in the final decisions concerning the plant.

ENVIRONMENTAL CONSIDERATIONS

Choice of Fuel

The fuel to be consumed is among the first decisions made when contemplating a new generating station, and even when additions to existing plants are proposed. The choice is not only predicated on its cost and projected future costs, but also on its consequences on the utilities position as a good citizen and good neighbor.

Earlier, the processes involved when different fuels are employed in a large part determine the economic selection of the fuel. These include costs incurred from the initial handling to the disposal of residue which may include smoke, radiation, etc. Often a more expensive fuel is chosen because of its greater acceptability by the community. Typically, oil is chosen over coal, and some oils producing less smoke and sulfur compounds are often specified or legislated. Indeed, this consideration was one of the reasons for the choice of nuclear fuel.

Emissions and Combustion

Two significant oxides of nitrogen caused by combustion are nitric oxide (NO) and nitrogen oxide (NO_2) commonly referred to as NO_x in environmental matters pertaining air quality. They affect the environment through production of acid rain and smog resulting in defoliation, direct and indirect deleterious effect on health, and reduction of the ozone layer (greenhouse effect). The majority of NO_x emissions are caused by combustion of fossil fuels. Combustion of coal, oil and gas in stationary devices such as power stations and in transportation movers such as trains and trucks generate about 40% of total world NO_x emissions. Electric generating stations are estimated to produce about 25% of this total, only part of which is due to using coal as fuel. From this, a conclusion can be drawn that emissions of N_2O are mostly from natural sources (i.e. forest fires) and not human activity—but the latter has been increasing over the last several decades.

Considering past and present environmental damage, and suggested future trends, governments are now recognizing the need for controls resulting in NO_x reduction. In 1970 and again in 1977 the United States Congress legislation—Clean Air Act—enacted standards for new construction of power plants requiring the installation of pollution controls that result in "the lowest achievable emission rate" or "the best available control technology" depending on whether the area did not meet federal standards or where the area is meeting federal standards. Standards may differ from state-to-state, but generally "best available control technology" for new coal plants is:

- selective catalytic reduction (SCR) which will achieve an emission of 15 pounds (NO_x) per MMBtu of coal burned

- scrubbers for SO_2 which will achieve emission rates of less than .3 pounds per MMBtu of coal burned

Regulations in general at present do not require older coal plants to meet these current standards. These plants could emit over .5 pounds of NO_x per MMBtu and 6 pounds of SO_2 per MMBtu of coal burned.

Other Air Pollutants

The incomplete combustion of fossil fuel may also produce significant amounts of carbon monoxide. Again, under certain conditions, much of this carbon monoxide is converted to carbon dioxide and carbonic acid spray, both of which are not harmful or objectionable.

Other pollutants are the gases, oxides of nitrogen, hydrocarbons and ozone. There is also some solid matter consisting of fly ash, soot and smoke.

Pollution Control

With present technology, electrical energy cannot be generated economically from "natural" fuels economically without at the same time producing some undesirable residue. Earlier, the pollutants contained in smoke have been discussed and their effect on surrounding areas minimized through the installation of scrubbers, dust collectors and precipitators, and the erection of stacks higher than would otherwise be minimally required.

The high stacks serve to dissipate both the sulphur and carbon compounds over wide areas into the upper atmosphere so that their effect at ground level is negligible.

Sudden load increases of load may call for a rapid increase in steam supply. This, in turn, generally results in more incomplete combustion of the fuel producing a greater amount of smoke, sometimes in voluminous quantities. This sudden increase in demands for electricity may be brought about by foreseen or unforeseen storms or other weather conditions, from special or unanticipated events and situations, or from the sudden interruption of other sources which a generating station may be called upon to assume. The "spinning reserve" generators help to mitigate such occurrences, as does switching of transmission lines where available.

Other Environmental Problems

Another potential environmental problem concerns the use of water to cool the steam in conductors; the water may be from bodies of water. The increase in temperature is usually limited by design to a maximum of about 30°F. This may affect the ecology of the body of water in question. The effect of thermal addition on marine life in the vicinity is monitored; should it be found deleterious, generating units may be shut down or other steps taken to remedy the condition. On the other hand, the heat added to certain bodies of water may actually cause marine life to flourish in the warm water. This is especially true of shell fish.

Another possible source of pollution is the handling and storage of oil used at the generating plant. Where this may be delivered by tanker or barge, expensive off-shore fuel terminals may be constructed, avoiding deliveries at the shore, and minimizing marine traffic and avoiding waters normally used by fishermen and pleasure boaters. On shore, the storage tanks are surrounded by protective dikes designed to contain the full contents of the tanks they surround in the event of accident. Often, they are also landscaped.

Nuclear Plants

In nuclear plants, very minute quantities of radioactive matter may escape from the fuel or are formed outside of it. Such matter is collected by waste purification systems and shipped away. The remainder, an exceedingly small quantity, may be released to the environment (air or water) on a controlled basis and in compliance with rigorous regulations established and monitored by the Nuclear Regulatory Commission. In general, this quantity may be so small that a large plant routinely operated for some 30 years may release a total of something less than a third of an ounce of radioactive matter to the environment over its entire lifetime of service. It must be remembered that our environment is naturally mildly radioactive, and that the amount of radioactivity released to the environment is only a very small proportion of the "background" radioactivity, so small that it is difficult to measure any increase in radioactivity above the background levels. See Figure 4-36 (a), (b), (c) and (d) in Chapter 4.

Noise and Appearance

Another important impact of generating stations on the environment involve both noise and appearance. Generally, the plants are located sufficiently distant from the more densely inhabited areas so that the noise emanating should not be a nuisance. In some instances, however, usually because of prevailing winds, the noise may be objectionable; in these instances, acoustic barriers may be erected at strategic locations.

In addition, appearances (or esthetics) are given much attention. Trees, shrubs and other plants, in addition to well kept lawns, are included in landscaping to beautify the plant site. Well designed buildings, with plenty of open space, beautifully landscaped grounds, and off-street or hidden from view parking space, enhance the scenic appearance locally and constitute an attempt at integrating facilities with the environment.

Where coal piles may exist, they may be enclosed within a suitably high wall not only to screen the coal from view, but to help in preventing wind from blowing dust outside the storage area; the wall and surrounding area is also landscaped. Oil and gas tanks may be placed underground or landscaped to improve appearances.

More and more utilities are committed to the protection of the quality of the environment in the areas in which they serve and, in every feasible way, implement that commitment.

Public Relations

As in any well run business enterprise, the interests of the consumer are paramount. This applies not only to the quality of the product, but the manner in which it is served. In addition to direct communications between the utility and its consumers, there are also numerous federal, state and local units representing the public interest and its operations.

In addition, employees at the generating plant take part in community affairs, often sponsoring, initiating and participating in projects concerning the education, health, finance, recreation and other local programs.

In this regard, it may be noted such plants contribute greatly to the economy and well being of the areas in which they are located. Not only do they provide many jobs to the local citizens, but are generally the largest taxpayer; school boards, police and fire departments, water and sewer services, all find their budgets well provided for, while residents also enjoy relief in their individual tax burdens.

CONSERVATION

Through the selection of fuels, the operation of the most efficient units for base loads and less efficient units for short time peak loads, much fuel is conserved. By interconnecting generating plants into a pool or grid, further fuel economies are realized.

While these operations apply to the utility owned and operated plants, further conservation of fuels takes place by including generation of electric power by the consumers, known as cogeneration.

Where consumers utilize fuel in producing steam, heat, or other outputs in their undertakings, the excess outputs are generally exhausted to the atmosphere as waste. Harnessing this waste energy to generate electricity, and connecting them to the utility's electrical delivery systems, permits the utility to use less fuel to produce less electrical energy. In some states, laws make it mandatory for the utility to seek out and connect to its system such electric capacity as can be made available from the waste heat products of its consumers. The problems associated with such connections will be discussed in more detail in the description and operation of electric delivery systems. See Figure 5-22 (b) in Chapter 5.

But not only fuel is conserved by cogeneration. The capacity available from consumers can act to defer construction of additional generating facilities by the utility, permitting it to conserve its capital outlays and the associated expense of raising the money required all of these factors not only have influence on consumers bills, but also influence the utility's relations with its consumers and with the several bodies representing the public interest. There may be a few drawbacks to cogeneration, however. The rate paid by the utility to the cogenerator for a unit of electrical energy may be greater than that at which the utility can produce it. Problems of protective devices coordinating with those of the utility may also prove serious. Figure 8-1.

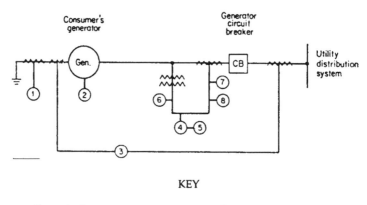

KEY

1. Ground relay
2. Generator governor
3. Differential protective relay
4. Frequency meter and relay

5. Synchronizing device
6. Undervoltage relay
7. Time-overcurrent relay
8. Directional power relay

Figure 8-1. One-line diagram showing protection relaying for consumer co-generation unit.

Chapter 9

Green Power

T he pollution of the atmosphere caused by the emission of substances detrimental to animal and plant life and to the integrity of structures has led to the investigation of other means of producing electric power, ones that do not contribute to the pollution. These have been generally referred to as Green Power or sometimes as renewable energy sources.

The emissions from fossil fuel fired plants include gases such as oxides of carbon, sulfur, nitrogen and mercury. Coals produce the greatest amount, with oil and natural gas less in that order. While the emissions are screened to eliminate as much as they can the solid particles, the escaping gases combine with the moisture in the air (ram) to produce destructive carbonic, sulfuric, and nitric acids, especially the latter two; mercury is usually associated with coal and may be found in small amounts in its combustion residue. These gaseous emissions carry with them 'waste' heat that also impacts on the generating plants efficiency as well as on global heating. Hence the search for and development of Green Power systems.

Included among the more promising systems are wind, fuel cells, solar, hydro, nuclear, geothermal, micro-turbines, and bagasse (the burning of some vegetation).

WIND MILLS

These consist of propeller blades turning generators located at the top of pole and tower supports (Figures 9-1 and 9-2)). Capacities of the generators, depending on wind conditions, may vary from a fraction of a megawatt to several megawatts. The outputs generally are varying with the wind velocity. The electric output increasing as the cube with the wind velocity, i.e. double wind velocity will produce 3 or 8 times as much output. As velocity tends to increase with height above ground, the higher the mounting, the greater the velocity and electric output, Economics plays a great part. The higher the structure, the greater its cost, and this must be balanced against the greater costs of the generators, until a satisfactory ratio of costs is reached. Also the length of the blades of the propellers must be taken into account, the longer the blades or blades (usually two or three) the greater the electric output, although there is a practical mechanical limit as to their size.

As winds are variable in velocity and direction, propeller blade angles or change in pitch may be varied and the whole turret or nacelle mounted on the top of the structure may be rotated to keep the propeller facing into the wind (known as the yaw). The extent of these maneuvers again influenced by economics.

As the electrical output sought may require more than one such unit, care must be taken in situating these units so that the air flow disturbed by one operating unit does not affect its companion units.

Electrically, the output must conform to strict limits of voltage and frequency to interconnect with other units or the utility system, if desired. Hence the speed and pitch of the propeller blades must also be controlled so that if the wind should

GE Wind Energy's 3.6 megawatt wind turbine is one of the largest proto-types ever erected. Larger wind turbines are more efficient and cost effec-tive.

increase above or drop below certain limits, the unit is taken out of service (and, of course, when there is no wind).

As winds tend to be greater in open areas, such as off shore or level agricultural areas, costs of associated transmission and distribution systems (if any) involving transformer, switching and protective circuitry must be considered as well as associated maintenance expenses. On the other hand, areas where such wind farms may be located may still be used for their original purposes or other productive uses.

Practically all wind power installations employ propeller driven horizontal shafts. Air turbines driving vertical shafts are sometimes used where appearance and bird safety may be considered (Figures 9-3 and 9-4).

Inside the Wind Turbine

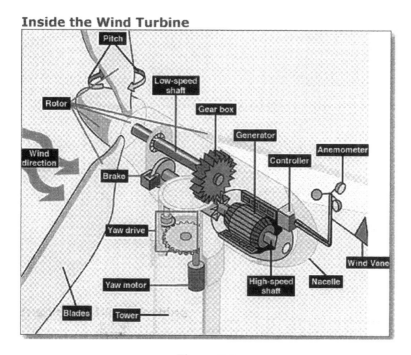

Figure 9-2.

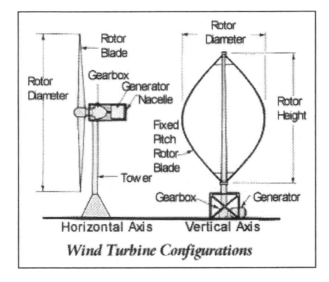

Figure 9-3

THE FUEL CELL

The fuel cell is essentially an electrochemical battery in which the material associated with energy is consumed, whereas in the fuel cell that material or fuel is continually replaced. The supply of such fuel constitutes, in effect, a storage of energy ready to be used on demand. The cell usually consists of an anode and a cathode, the materials in contact with an electrolyte that is most often a fluid. Any one, or all of these materials may be consumed and replaced.

Figure 9-4 (Courtesy LIPA)

In the fuel cell, the fuel reacts electrochemically, releasing ions that travel via one of the electrodes to the external circuit to the connected electric load. (Doing useful work) The ions return via the other electrode to combine with oxygen ions (from air) to oxidize the opposite polarity fuel ions that traveled there via the electrolyte. Although the ions can travel in either direction, for simplicity they are usually portrayed as leaving the anode and returning via the cathode. At both electrodes, the chemical reactions produce heat and depleted materials to be disposed. (Refer to Figure 9-5)

The types of fuel cells are most often described by the electrolytes used, which may be gaseous, fluid, or solid. Currently five types may be identified:

As heat is generated in the action of these cells, and as some escaping gases may evolve, ventilation and some form of cooling may be necessary. The stacking of such cells to attain desired outputs may make such measures imperative.

Because of its availability twenty four hours a day, and its potential to be located at or near its centers of utilization, this unit may not need associated transmission and distribution systems. However, should it be desirable to do so, its output as direct current may need to be converted to alternating current through the use of "inverters" or other devices. The addition of these rectifying devices with their control and protective circuits (and including their maintenance costs) detract from their overall economic advantages.

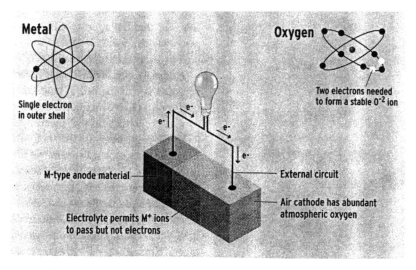

In this generic fuel cell, anode material M (to suggest a metal) has one electron in its outermost electron shell, which it gives up fairly easily, forming an M^+ ion. The oxygen at the cathode tends to acquire the two electrons needed to fill its outer shell.

Because the electrolyte blocks electron flow, the electrons liberated at the anode must pass through the external circuit where they can do useful work. The M^+ ions can cross the electrolyte, which lets them react with the O^{-2} ions to form M_2O, the cell's waste product.

Figure 9-5. Fuel Cell Theory (*Courtesy IEEE Spectrum*)

1. **Proton Exchange Membrane**—A polymer that is a proton conductor and water. A hydrogen rich gas is the fuel, reacting with a catalyst electrode, often platinum. Operating temperature, limited by the polymer, is usually less than 120°C.

2. **Alkaline Fuel Cell**—The electrolyte is a concentrated sodium or potassium hydroxide, retained in a matrix, and many kinds of metals and metal oxides used as electrocatalysts. Fuel supply may include materials that do not react with the alkaline electrolyte, except for hydrogen. Operating temperature may react 250°C.

3. **Phosphoric Fuel Cell**—Concentrated acid is used for the electrolyte retained in a matrix (of silicon carbide). The electrocatalyst in both electrodes is usually platinum. Operating temperature is in the 150-220°C range.

4.   **Molten Carbonate Fuel Cell**—The electrolyte is a combination of alkali carbonates of sodium and potassium, retained in a ceramic matrix. The alkali carbonates form a molten conductive salt with carbonate ions providing the ionic conduction. Operating at a high temperature of some 600-700°C, nickel anodes and nickel oxide cathodes are adequate to promote reaction.

5.   **Solid Oxide Fuel Cell**—The electrolyte may be gelled potassium hydroxide or strontium oxide, while the fuel may be aluminum or zinc. The reaction liberates oxygen ions as carriers of the electric charge. The metallic fuel is connected to the anode of cobalt or nickel oxide and the cathode of strontium clad metal. Residue may consist of zinc oxide that may be reconstituted into zinc in the cell. Operating temperature may range from 800' to over 1000°C. (Refer to Figures 9-6 and 9-7)

While gaseous hydrogen as a fuel has many good features because of its high reactivity with proper catalysts, its availability and its simplicity of operation, the large volume necessary for practical purposes require containers to withstand the high pressures and extremely low temperatures to maintain its liquidity for handling. Storage facilities may be replaced with external generators of hydrogen gas, the installation of which detract from its economic advantages. All of which allow other fuels to be competitive.

Fuel cells generally consist of an electrolyte sandwiched between two thin plates acting as electrodes. The fuel passes over the anode and air (oxygen) over the cathode. The modular construction lends itself to their assemblage in combinations suitable to the job in hand. (Refer to Figure 9-8)

The electrolyte functions to conduct ionic charges between the electrodes, thereby completing the electric circuit. It transports dissolved reactants from the anode to the electrodes, and provides a physical barrier preventing the mixing of fuel with oxidant materials.

The voltage developed by fuel cells generally is one or two volts. Fuel cells may thus be connected in series to deliver the voltage required for a particular application, and stacks may be connected in multiple or parallel to provide the desired capacity.

The characteristics of fuel cells that make them suitable for Distributed Generation, particularly as electrical energy without the

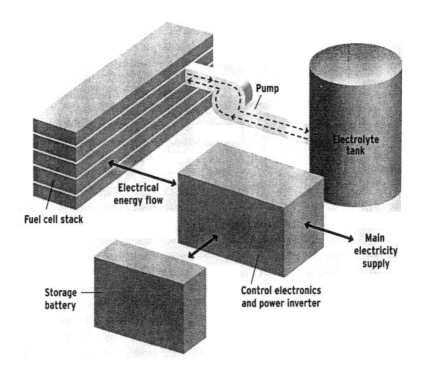

Too Corrosive to Be in Constant Touch

Aluminum-Power's aluminum-air system stores its corrosive electrolyte in a tank away from the stack when the cell is not operating. The electrolyte is pumped into the stack to start the cell operating, and out to shut the stack down. Keeping the electrolyte separate from the stack eliminates parasitic corrosion, thereby ensuring that the system maintains its capacity during long idle periods.

Figure 9-6. Handling of Aluminum, Air Fuel Cell (*Courtesy IEEE Spectrum*)

combustion of fuel. They are hence environmentally attractive, have high-energy conversion efficiency, flexibility in fuel used, rapid load responses, low chemical and noise pollution, modular design. And low investment costs compared to other forms of generation.

SOLAR POWER

Solar energy may be directly converted to electrical energy by concentrating its rays on semi-conductor materials that then produce

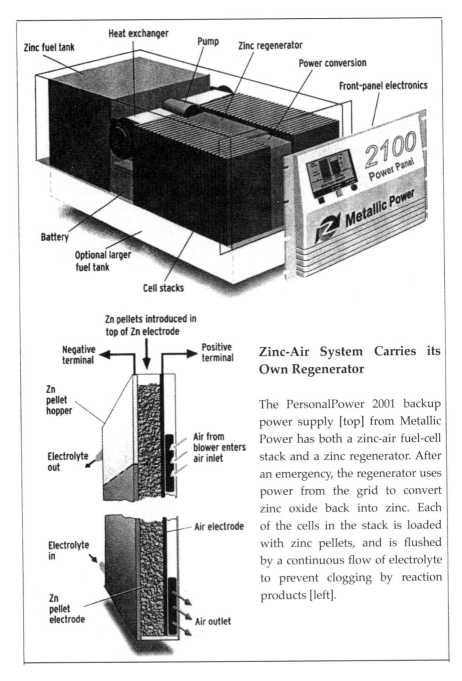

Zinc-Air System Carries its Own Regenerator

The PersonalPower 2001 backup power supply [top] from Metallic Power has both a zinc-air fuel-cell stack and a zinc regenerator. After an emergency, the regenerator uses power from the grid to convert zinc oxide back into zinc. Each of the cells in the stack is loaded with zinc pellets, and is flushed by a continuous flow of electrolyte to prevent clogging by reaction products [left].

Figure 9-7. Handling of Zinc-Air Fuel Cell (*Courtesy IEEE Spectrum*)

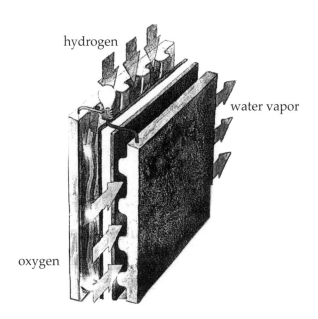

hydrogen

water vapor

oxygen

Figure 9-8. Fuel Cell Unit Showing Construction (*Courtesy IEEE Spectrum*)

electrical power or energy. Selenium is the material generally employed. Trays of crystalline selenium that required periodic raking have been replaced by ribbons consisting of extremely thin layers of selenium, indium and copper, applied to glass that is exposed to the sun's rays. Periodic replacement of portions of the ribbon exposed to the sun may be required. The electricity produced, like the fuel cell is direct current, and all the means to connect them to alternating current utility systems mentioned for fuel cells apply to the solar panels, including some form of ventilation. Depending on the areas of the panels (Figure 9-9) and the reflecting means (Figure 9-10) of concentrating the sun's rays, outputs of electric power may range for fractions of a megawatt to several megawatts. Unlike the fuel cell, the unit is tied to one location and may not generate electric power when the sun is not shining for whatever reason.

The solar system may also contribute to the production of electric power by heating, through heat exchangers, the water that is used in boilers that generate steam to turn the turbines that turn the electric generators, or to provide hot water or space heating. If installed in areas of harsh winters, precautions should be taken to keep the water in the exchangers and associated plumbing from freezing, usually by the rapid circulation of the water.

Use of the sunlight may also be made through its transmission via fiber optic conductors that also act as the source of lighting. This use is obviously restricted in application.

Figure 9-9. A 28-kW photovoltaic (PV) system is integrated into the roof to prpovide solar electricity for the building. Courtesy DOE

Figure 9-10. Parabolic troughs make up this concentrating solar power system. Courtesy of DOE

HYDROPOWER

While hydro power provides a clean means of producing electricity, it is usually an expensive means and is almost always associated with collateral usages, such as providing potable water, flood control, transportation links, recreational facilities, and other local benefits. Notable examples are Hoover Dam and Bonneville Dam. Smaller installations (Figures 9-11 and 9-12), providing some or all of these advantages, may prove desirable and so are included here as sources of Green Power. Like wind and solar power, hydro plants are subject to the whims of nature, depending on the rate of flow of available water that may be zero or small during periods of draught.

HYDRO-WATER WHEEL

All water power is derived from the pressure or force exerted by falling through a given distance (height) or "head." While water wheels turn the generators to produce electricity, the whole hydro system must

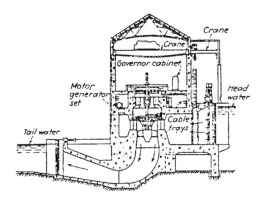

Figure 9-11. Cross-section of typical low-head concrete spiral-casing setting. Turbine at 90 rpm under 23-ft head. (Courtesy American Gas and Electric Co.)

Figure 9-12. This fish ladder on the Ice Harbor Dam on the lower Snake River provides safe passage for migrating fish. Courtesy DOE.

be considered as the prime mover. The elements of a hydro electric plant include:

1. *Dams* which impound and store the water and may be of earth, rock, concrete, or a combination of any or all of these.

2. *Intakes* which lead the water to the plant and which may consist of canals, flumes, pipelines, pressure tunnels, or a combination of some of these.

3. *Penstocks* which generally consist of one or more pipes connected in parallel and in which long penstocks must have sufficient diameter to prevent the condition known as "water hammer." Water hammer is caused by surges in pressure and velocity reduction in the pipes when more or less sudden changes occur in the smooth and uniform flow of water.

4. *Valves and gates* which control the flow of water to the water wheel—sometimes called a water turbine—and which may be of different types and configurations.

5. *Racks* which protect the water wheels against ice, trash, and other debris and which are essentially steel screens whose spacing may vary with location.

6. *Water Wheels* which are attached to the generator rotors and which may be of two general types: reaction or impulse wheels, also sometimes referred to as Francis wheels. The reaction wheel utilizes the water pressure and the reactive force on the curved blades which tend to change its direction; these are essentially propellers with fixed or adjustable blades. The impulse wheel, sometimes referred to as the Pelton wheel, utilizes the velocity and impact of a jet of water directed against buckets on the rim of the wheel; these are generally limited to very small plants. The reaction wheels are best adapted to relatively low heads and large quantities of water; the impulse wheels to high heads and small quantities of water.

7. *Tail Races* which carry away or discharge of the "spent" water, with a minimum sacrifice of head, to the body of water or stream below the dam structure and which may consist of concrete or steel pipes, tunnels, or other passageways.

Although hydro plants consume no fuel for generating power, and are very efficient in the 90 percent or better range, they are not only dependent on the availability of water, but are comparatively extremely costly. Costs of operation and maintenance, however, are generally low. In many cases the cost of the project may be shared with other purposes, such as, flood control, water supply and conservation, recreation, etc. Because of the possibility of inadequate or curtailment of water availability, such plants are often supplemented by other (fuel-fired) generating and transmission facilities.

Other applications of hydro power include so-called pumped storage installations and those associated with tidal flows. As the name implies, water is pumped uphill into reservoirs, usually lakes, during off peak hours, by the generators acting as motors driving the water wheels as pumps. During peak hours, the water is loosed (through penstocks) to drive the water wheels turning the generators to produce electricity.

In the case of tidal flow, advantage is taken of high tide s filling reservoirs that may be emptied at periods of low tide to turn water wheels that turn generators to produce electricity. The variations between high and low tides must be great enough (such as at the Bay of Fundy) for tidal projects to be feasible.

NUCLEAR POWER

Units of nuclear power generation must generally be of large capacity to justify the huge investments necessary for their installation. Their operation in some derail is described in Chapter 2, pages 42 to 50) and are included here as producers of Green Power.

GEOTHERMAL POWER

Geothermal power may generally be produced from springs found in almost every part of the earth, a concentration of them in specific areas lend themselves to power generation. Springs constitute the normal flow of water from the ground or from rocks, representing an outlet for the water that accumulates in permeable rock strata underground. When the water, because of the geological structures of the strata, issues under pressure (artesian), it may be used directly to drive turbines turning generators to produce electricity. More often the water at relatively lower temperatures, approximately 400°F, issues (geyser) is circulated through heat exchangers to preheat water that ultimately is changed to steam for driving turbines turning generators. So called "hot springs" occur when the water issues from great depths or is heated from nearby volcano.

Geothermal power plants may be classified by their mode of operation. When steam is used directly, it is termed a Dry Steam Power Plant (Figure 9-13). When hot water is sprayed in a tank at lower pressure than the water, it may cause the water to turn into steam, or 'flash' that drives a turbine turning a generator, and is termed a Flash Steam Power Plant (Figure 9-14). When the hot water is used to preheat water to be turned into steam in a boiler, using heat exchangers for this purpose, the plant is known as a Binary Cycle Power Plant; most geothermal power plants are of the binary cycle type (Figure 9-15).

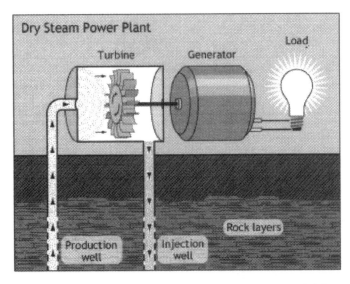

Figure 9-13. Dry Steam Power Plants. Courtesy DOE

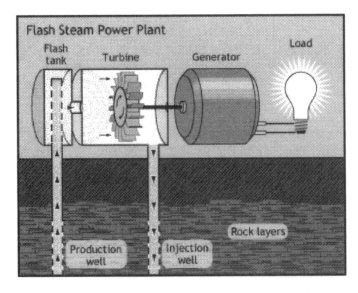

Figure 9-14. Flash Steam Power Plants. Courtesy DOE

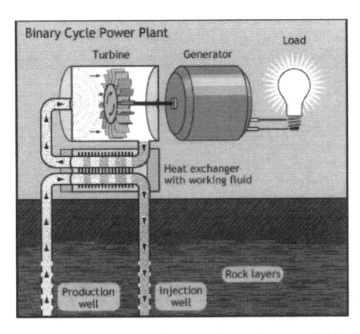

Figure 9-14. Binary -cycle Power Plants. Courtesy DOE

MICRO TURBINES

These are small capacity turbines driven by natural gas, and are described in Chapter 5 (pages 191 through 193) Although the exhaust from these turbines contain pollutants, never the less because such emissions are relatively low compared to larger coal and oil fired installations, they are included as Green Power plants.

BIOMASS

Also known as Bagasse, this material consists of the residues of forest and agriculture together with (municipal) solid waste, greenhouse and landfill gases. It constitutes the greatest potential source of production of Green Energy, even though it plays a relatively modest part in the production of electricity where it is a part of the output of the vast chemical industries. Its part in the direct production of electricity consists in its taking the place of coal in boilers providing steam for turbines driving electric generators; this may be the direct burning of it, or its burning with coal, diluting the effects of the latter. The same disadvantages, though to a lesser extent, apply with the same measures of mitigation associated with coal.

Its better application relies on its gasification and liquidation and used as the raw material for the chemical industries, essentially substituting bagasse for coal, oil and gas in the petrochemical field. The production of ethanol, bio-diesel fuels and gases including methane and other flammable gases can be used as 'cleaner' replacements for fossil fuels, all easier to handle than coal. This raw material is changed by chemical processing into valuable products that include many types of polymers and plastics, synthetic fuels, wood, and artificial stones, animal feed, a number of drugs and chemicals, and a host of other products, including industrial process heat and steam, all of which make for conservation of natural resources. The flammable gases produced may be cleaned and filtered to remove problem chemical compounds

Index